AF545183

Care and Repair of Hulls

By the same author

Complete Amateur Boat Building
(John Murray)

Boat Repairs and Conversions
(John Murray)

Boat Maintenance
(Kaye and Ward)

Building Chine Boats
(Yachting Monthly)

Lifeboat into Yacht
(Yachting Monthly)

Michael Verney

Care and Repair of Hulls

ADLARD COLES LIMITED
GRANADA PUBLISHING
London Toronto Sydney New York

Published by Granada Publishing in
Adlard Coles Limited, 1979

Granada Publishing Limited
Frogmore, St Albans, Herts AL2 2NF
and
3 Upper James Street, London W1R 4BP
1221 Avenue of the Americas, New York, NY 10020 USA
117 York Street, Sydney, NSW 2000, Australia
100 Skyway Avenue, Toronto, Ontario, Canada M9W 3A6
110 Northpark Centre, 2193, Johannesburg, South Africa
CML Centre, Queen and Wyndham, Auckland 1, New Zealand

ISBN 0 229 11577 2

Printed in Great Britain by
Fletcher & Son Ltd, Norwich

Contents

one

Ownership

The reader of this book does not need to be sole owner – or even part owner – of a boat. From the very start in boating, if you wish to become a sought-after crew member, learn the basics of boat care and repair. Too many people enjoy racing or cruising, but are useless when it comes to maintaining the boat in good shape.

A surprising number of newcomers first take to the water in a new boat chosen from brochures or from the exhibits at a Boat Show. This may prove a success, but how much wiser to get the feel of it first by joining a club and assisting other owners for a while. If racing is the main attraction, choosing the right boat for local conditions – or to compete with the established classes – is often essential.

Chartering

When cruising invitations are not forthcoming, chartering (hiring) a boat during the first season or two helps all concerned to gain experience. This may have to be on inland waters. The owners of charter boats are naturally reluctant to release these into tidal waters unless a competent skipper is on board. For a family including children to get a taste of life afloat without embarking on ownership, chartering is often the best choice.

Only the most basic hull care and repair know-how is normally required when chartering, but some attention to engine and equipment is sure to be necessary, while thorough cleaning before handing back at the end of the holiday is considered courteous, even if a dirty ship does not bother you when on board!

Syndicates

Few boat owners manage to spend as much time afloat as they would wish, so that thoughts about putting the money involved to better use may arise. Sharing all costs between the members of a syndicate then proves tempting and, although tales of unsuccessful syndicate ownership are legion, it can work well.

Favourable conditions often exist when two people with approximately equal amounts of leisure time form a sailing dinghy partnership, possibly alternating as helmsman and crew. Similarly, six or eight people could own an ocean racer and sail as a team in the maximum possible number of races each season. Syndicates owning high speed powerboats sometimes work well. Should a disagreement occur, one person's share should not be too difficult to sell. To cut costs and ensure safety, all members of a syndicate need some knowledge of boat care and repair.

The Right Boat

Few yachtsmen seem to stay satisfied with a boat for long – this helps to keep the brokerage business thriving. However, the occasional owner keeps the same boat for thirty years or more and becomes very attached to her.

Short-term owners tend to sell just before any major maintenance work becomes necessary. Long-term owners tend to be perfectionists, attempting to keep the old ship looking like brand new at all times.

The impecunious can (with careful buying and selling) make a profit by changing boats frequently, increasing their knowledge greatly in the process.

The wide beam of multihulls leads to costly mooring and dinghy pen space. For easy trailing and stowage, some multis have adjustable or removable beams. Light construction coupled with high speed renders multis vulnerable to costly collision damage – beginners beware!

Most small powerboats tend to be fast nowadays and may not handle safely at the restricted speed of perhaps five or six knots authorized on most inland waters. High-speed craft are extravagant on fuel and can sink without trace if they collide with a half-submerged object.

The Right Engine

Diesel fuel is hard to beat for economy and is therefore the obvious choice when long-distance cruising under power is anticipated. On account of the greater thirst and fire risk of a comparable petrol (gasoline) fuelled boat, the secondhand value is lower, though the engines are quieter. Such a craft could well fit the pocket of the prospective owner who intends to make short cruises and has sound mechanical knowledge.

Much the same applies to the auxiliary motors of sailing craft, but the powerful motor/sailer type invariably follows big power yacht practice by adopting diesel fuel. Engines without spark ignition are an obvious advantage in the dank dungeons of a boat's bilges!

Living Aboard

Except for trailer-boats, ownership is an expensive pastime, but enthusiasts will tell you that there is no more worthwhile way of spending money. Owning a cruising boat certainly saves hotel bills and other conventional leisure time expenses. Some owners pay for all the boat's running costs by chartering her out for part of the season.

Turning a boat into permanent living quarters pays off handsomely and the savings enable one to afford a big enough craft to have good accommodation, perhaps with a storeroom and workshop on shore near by; the idea generally suits retired couples best.

Under this heading, houseboats may also be considered. These are often defunct yachts or custom-built pontoon caravans which rarely move. One of these can make a comfortable home, particularly when your cruiser, runabout, or racing boat is moored alongside. Houseboats need care and repair in similar fashion to their more active counterparts.

Climatic Influence

Timber construction ensures comfort in most climates, giving a more constant internal temperature than plastics (except foam sandwich), metal, or ferrocement. Except in the tropics, some sort of cabin heating is necessary for winter cruising. Further towards the Arctic, you may need an air bubbler under the boat to prevent ice forming around her. Air-cooled engines prove their advantages while fresh water cannot be stored in unheated tanks.

Cabin craft keep cooler in summer with white or pale coloured topsides, but beware of glare from decks which are too light: scrubbed teak is ideal.

Traditional wooden boats do not take kindly to conditions in some dinghy pens and boat parks which quickly turn into sun traps in fine weather. Plywood hulls stand such conditions quite well, but light alloys and plastics cope best of all.

In low latitudes, moving along on sea, lake, or river, can be the coolest place. In port, however, a boat should have large ventilation openings with wind scoops, also raised awnings or canopies, preferably equipped to cover the whole ship. Without such facilities, craft designed for British waters are rarely suitable for a Mediterranean summer. Similarly, those designed for New England waters may not prove suitable in Florida or the Caribbean Islands.

Hours of daylight are shorter in the tropics and equipment for night navigation needs to be efficient. A deck-wash hose system is of great benefit for keeping the decks cool in hot weather.

Choice of Materials

Although timber is still the most popular material for amateur boat builders, few professionals use it nowadays for hull construction. Plastics dominate the scene at present for boats of small and medium size, while steel is unbeatable for the big custom-built craft and competitive in the medium range. Light alloys are used for all sizes (particularly in North America) and their low weight also makes them attractive for the superstructures of big boats with hulls of other materials. Ferrocement is the cheapest material for building boats in the 40 ft (12 m) to 80 ft (24 m) range. Although used with success by both amateur and professional boatbuilders all over the world, this material is not universally accepted amongst boat owners.

When choosing a boat, initial cost may be more important than the various advantages and disadvantages of different materials. It often happens that a low price, whether due to age, design, or method of construction, means higher upkeep costs, but many people prefer to distribute their money in that way.

Craft built without much wood (plate 1) have a veritable bath-tub appearance. In general, the more timber used for decks, coam-

Plate 1. Modern cockpit without hardwood trim.

ings, seats, joinery and trim, the more annual maintenance she is likely to require, but the more pleasing she will be in appearance and use (see plate 2).

Maintenance Requirements

Provided the correct grade of alloy has been used, a polished aluminium boat which is dry-sailed (i.e. kept ashore when not in use) needs no painting and little cleaning. The reflective

Plate 2. Traditional cockpit in varnished wood.

surface helps to prevent extremes of temperature when such a boat is decked over, but few people like the external appearance.

If you give plastics the same sort of beauty treatment as the average car which is parked near a sea-wall for five years, the surfaces are certain to become dull and blotchy in both cases. White lasts better than colours, but sooner or later painting is necessary to regain the original smartness.

A foam sandwich plastics hull does not normally have a pigmented gel coat, so paint is applied from new. The maintenance requirements are therefore little different from timber, steel, or ferro-cement. Similarly, all types tend to use varnished wood trim which is bound to need care and attention.

Most boats kept afloat need antifouling paint under water. Curiously enough, old-fashioned wooden hulls with copper sheathing need no bottom paint. This makes them attractive propositions for extended ocean cruising where facilities for scrubbing and painting may be scarce or expensive.

Rigs and mechanical parts need similar maintenance work irrespective of hull material. Galvanized rigging needs more frequent renewal than stainless steel, while varnished or painted wooden spars require much more upkeep than their metal counterparts.

Tools

On small open boats, a knife may be the only tool carried. Even the most ham-fisted owner should carry a good tool-kit on any craft with mechanical equipment on board. A handy crew member or visitor may be able to fix an important running repair provided the necessary tools are available. It would indeed be galling to spurn such an opportunity, for cruising boats spend much of their time remote from shore facilities.

What tools do you need? Good question! A kit is usually built up gradually over several years. Only the deep sea wanderer is likely to carry carpentry tools, a hatchet and bolt croppers, but all cruisers need basic engine tools, plus various hammers, pliers, shackle key and screwdrivers, to enable simple rigging and electrical repairs to be tackled. The handyman is bound to keep adding tools as time goes by. Such an owner might enthuse over the acquisition of a portable vice. His wife is more likely to get excited over a battery operated vacuum cleaner!

Keep tools lightly oiled or stored near rust-inhibiting paper. Fit lanyards to all tools used on deck: they are so easily lost overboard.

Don't forget to carry a small but powerful magnet attached to a piece of copper wire to help in retrieving tools, nuts and other steel parts from the bilges, or indeed the sea bed, if they have been dropped overboard in shallow water. The magnet must be stored well away from the compass, of course.

two

Summer Care

Whether left afloat all year or laid up ashore for part of the time, all boats need in-season attention to retain a shipshape appearance, to ensure safe operation and to keep the interior smelling sweet. No harm may result in neglecting summer care, but whereas a few owners think that a dirty boat looks like an active one, the majority prefer to maintain start of season appearance at all times.

Few boats are entirely safe when sailed single-handed; some require a crew of three or more. Arranging to get under way is not always simple, but this does not worry the practical owner. Just being on board is sheer delight to him: doing the necessary chores is all part of his hobby.

Valeting

The paid hand on a wealthy owner's yacht is kept busy continuously when the owner is absent, just maintaining the status quo. Perhaps the owner would be happier if he had the time and inclination to do some of this work with his own hands!

Further down the scale, valet services operate around many anchorages, marinas and inland waterways so that owners with little spare time, or with no desire for just messing about in boats, can pay to ensure that the vessel is always kept shipshape. Such services vary and can include washing down deck and sails; airing and cleaning the accommodation; leathering off brightwork, windows, or portlights; washing topsides and scrubbing the bottom.

Topsides

Some ports and marinas are relatively clean – others are foul. Sea spray leaves salt streaks on pigmented topsides, but on a white finish, stains are most noticeable along the waterline, below deck scuppers and round exhaust outlets.

Most craft need topside cleaning at least twice during the season. Occasional dousing with a deckwash hose helps, especially if followed by mopping. Stains need close-up treatment with a bath cleaning powder or one of the special boating preparations stocked by chandlery stores.

Most of the surface can be reached from alongside a pontoon or slip, turning the boat if necessary. On a swinging mooring, useful work can be done from a dinghy, kept close in by a *guest warp* around the hull, to which the fore-and-aft dinghy painters are made fast, see fig. 1. If a painting raft can be borrowed, so much the better. This is really a job for swim suits and a hot day.

A long-handled domestic squeegee sponge works better on topsides than a deck mop. Beware of the pollution regulations in some marinas which may prohibit the use of detergents for cleaning a hull.

On small boats, a quick topside wash with the squeegee is possible by leaning over the

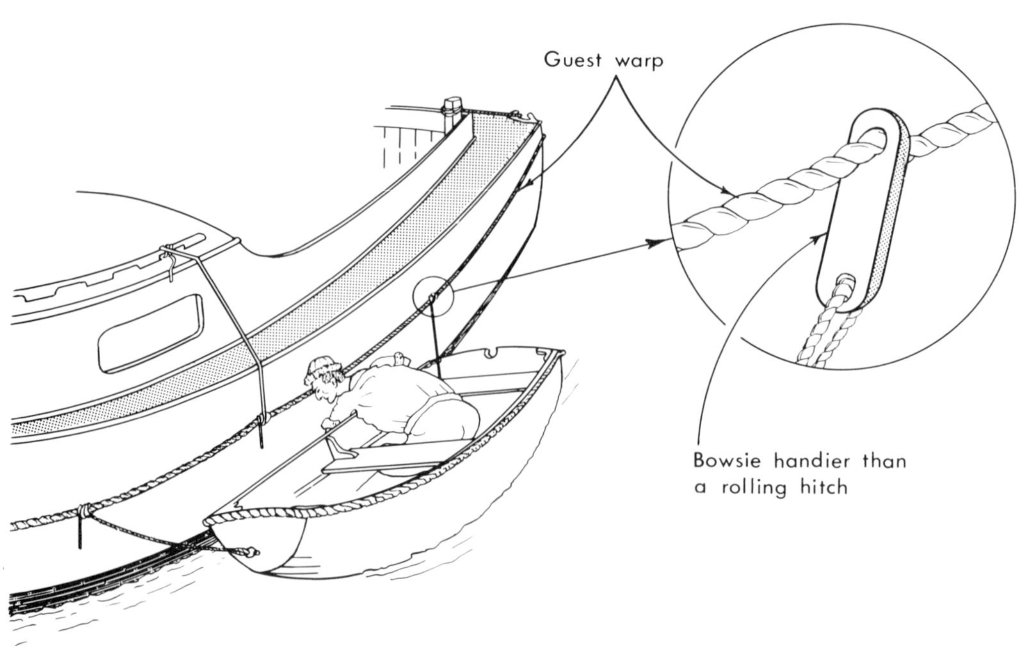

1. Securing the dinghy for a topside scrub.

deck rail, preferably with a helper to hold one's feet and to freshen the mop from a bucket of soapy water. Fresh water is always best, as brine leaves salty spots behind as it dries. A fresh-water hose from on shore speeds the job greatly.

Cleaning is simplified on plastics and bright aluminium if boat wax is applied at fitting-out time. It rarely proves practicable to apply more wax mid-season on craft afloat, though this is certainly possible on small boats which are kept in boat parks or dinghy pens. Wax will protect burnished surfaces painted with catalyst finishes, such as two-can polyure-

Plate 3. Getting rid of the salt after a race.

thane, but is useless on standard alkyd boat enamels and varnishes.

Particularly on wood and steel, paints are intended to protect as well as beautify. If any chipped or bubbled paint is discovered when cleaning down, the enthusiast will be wise to scrape away loose material and, when all is dry, apply a couple of coats of touch-up paint. Deep scores from anchor flukes or collisions usually need a skim of knifing stopper before painting – see Chapter 9.

Brightwork

Touching up weathered or damaged varnish work is usually more urgent than touching up paintwork. Speed is particularly essential on mahogany, which discolours rapidly on exposure; teak is far more tolerant. After a light rub down, re-varnishing will always achieve a good match. Other popular hardwoods such as iroko and afrormosia are not as critical as mahogany, but need more attention than teak. Edges of hatches and coamings are particularly vulnerable to chipping.

Brightwork can be kept in perfect condition if sponged down with fresh water once a week (or after a passage) then dried off with a wash leather. This is a time-consuming job on big expanses and is a much neglected maintenance chore.

Sails

The owners of sailing dinghies are generally conscientious types who wash down the whole boat with a fresh-water hose (plate 3) after every race in salt water. At the same time they usually wash down the sails, leaving them hoisted (with the weight on the luff) to dry before stowing in bags.

If the salt water can be washed from cruiser sails in this manner after a wet passage – or hoisted in the rain as soon as possible afterwards – the life of the sails will be extended and less mildew will form on them.

Self-adhesive nylon repair tape up to 2 in (50 mm) wide is ideal for the emergency repair of rips in spinnakers and other light sails. Alternatively, stick on strips of sailcloth both sides using Loctite cyanoacrylate glue. This method is also useful for repairing canvas covered canoes – see Chapter 4.

Windows

Cleaning windows and portlights is more important than brightwork, but the two jobs can be treated as one. Wheelhouse windows may need hourly cleaning under certain conditions of heat and salt spray.

Unbreakable windows of acrylic sheet (see Chapter 8) are popular on small coastal and lake cruisers. After some two years of conventional cleaning these may get scratched and remain looking dirty. This damage can be obviated by always cleaning such windows with metal polish (such as Brasso) followed by a polishing cloth or dry duster.

Bird Menace

Seabird droppings cause endless deck cleaning work in some marinas. This usually occurs while the owner is away. With no burgee at the masthead, a bird can perch happily at the truck. Hoisting a short bamboo cane in place of the burgee will eliminate the problem.

A permanent wind vane or anemometer at

the masthead prevents birds from perching, but on a ketch, schooner or yawl, they will soon find the other mast! If birds roost on the spreaders, a piece of thin shockcord rove as in fig. 2 will do the trick.

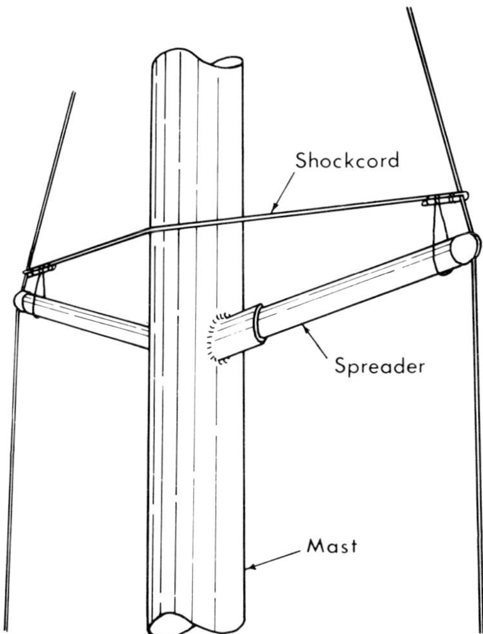

2. A simple seagull inhibitor for the spreaders.

Washing Decks

Sea water is a mild fungicide and is always used to wash traditional timber decks. If there is no power-operated deckwash hose, water must be scooped up using a bucket with a long rope attached to its handle. Should slight leakage occur on to the deck beams, hull frames, or planking, rot is less likely to start with sea water than fresh water. However, any visible leaks which could cause damp bedding should be sealed without delay (see Chapter 8).

Modern teak decking bonded to a plywood base does not need continuous wetting in hot weather to prevent seams from opening up, but washing is still advantageous to provide a cooling effect. An awning without side curtains cannot shade a deck completely against all angles of sunlight.

Bare teak decks are easily marked by dirty footprints, oil, food, fuel, and rust stains, but rubbing with a holystone and scrubbing with soap has been superseded by the special bleaches (such as Teakbrite) sold by marine stores. The teak veneered plywood with simulated seams used on runabouts is not usually walked upon and is often varnished, needing little but sponge and chamois treatment.

Foredeck

After weighing anchor or picking up a weedy mooring chain, the foredeck invariably needs dousing with copious bucketfuls of sea water. Once the mixture of seaweed and mud (with perhaps some globules of oil) has been allowed to dry, energetic scrubbing is needed to remove it.

On inland waters, the benefits of brine for deck washing are regrettably absent, but a dirty foredeck is bound to occur and even polluted canal water is probably better than polluted mud.

Anti-slip Surfaces

Smooth fibreglass decks are easy to keep clean, but large areas are normally covered with roughened strips which retain dirt and need scrubbing.

Canvas-covered wood decks are usually coated with anti-slip paint containing grit. This gets dirtier as the season progresses and it might not be a bad idea to apply a midsummer coat – at least to those parts such as side decks which are subject to the most wear.

The resilient compositions often used over metal or ferro decks rarely have a sand finish and are easy to clean with mop and scrubber. Trakmark and other synthetic embossed (or simulated wood) coverings scrub up cleanly until slightly worn, when special paint is available to rejuvenate them.

Scuppers

Dirt congregates in footrail or bulwark scuppers (freeing ports) and creates stains down the topsides after rain. The best tool for cleaning small scuppers is a bottle brush, obtainable in various sizes from pharmacies or hardware stores.

Without this, a deck swab or piece of old towel poked through and drawn to and fro will suffice, but it takes a long time. An ordinary scrubbing brush will deal with large scuppers. Remember to do this job before cleaning the topsides.

Rust Stains

On some steel hulls, particularly old ones or those which were not zinc sprayed after building, rust stains often appear mid-season around bulwark stanchions, hawse pipes, and where topsides have been scratched. Little time is required to stop this spreading by applying quick dabs of touch-up paint. Similar treatment on old timber-planked hulls may be necessary where once-galvanized rigging chain plates start to ooze rust stains down the topsides.

Below Water

When weed and barnacles start growing on a boat's bottom, her sailing performance suffers markedly, while a powerboat needs to burn more fuel (with higher engine revs and more noise) to obtain a normal cruising speed.

Small racing craft obviate this problem by storage ashore between races, when the unfouled bottoms can even be burnished, or re-coated with graphite paint or other slippery composition.

Antifouling

The best (and most expensive) antifouling paint can fail after two months in some waters, but the effectiveness varies greatly from coast to coast, even from port to port. When performance and economy are important, bottom cleaning is a big summer care problem. Most owners do nothing about it, but one should remember that in dangerous lee shore gale situations, or when engine trouble occurs, minimum hull friction can make all the difference between safety and stranding. When power is available, sonic vibrators fitted inside can prevent a lot of the usual fouling.

Inland water fouling is generally less severe than that in sea water, but warm fresh water can cause bad conditions. If a vessel normally moored in salt water is moved for a few days into a fresh-water basin or river, nearly all growth will die off. The reverse procedure also holds true. On some old canal systems the silt bed rubs off any paint applied to the lowest part of a hull.

Scrubbing

For skuba-diving enthusiasts, underwater scrubbing is simple if a line is rigged under the keel to hold on to. However, the job is exhausting and it may take a week of short spells to complete. Commercial ships are also troubled by fouling, but being under way almost continuously minimizes growth. Nowadays, mechanical scrubbers are set to work when a ship is in port.

Conventional antifouling paint is partially removed by scrubbing, so the process cannot be repeated many times without need for repainting. Hard racing type paints can be scrubbed many times without loss, but their antifouling properties are not quite so good.

Careening

Weed grows most prolifically just below the waterline, whereas barnacles usually prefer the dark regions around the keel. In waters with sufficient tidal range, it may be possible to let a boat dry out on a special grid or on a beach to permit scrubbing and repainting (plate 4). With only a small tidal range, she may be *careened* in shallow water by bowsing down with a masthead tackle made fast to a

Plate 4. Careening simplifies a low tide scrub.

buried anchor or a bollard on shore. However, note that most modern twin bilge keel shoal draft designs will not lie over in this manner, unless anchored fore and aft and careened in deep water.

Note also that having painted all reachable parts on one side, the paint may suffer some damage when she is careened on the next tide to tackle the other side.

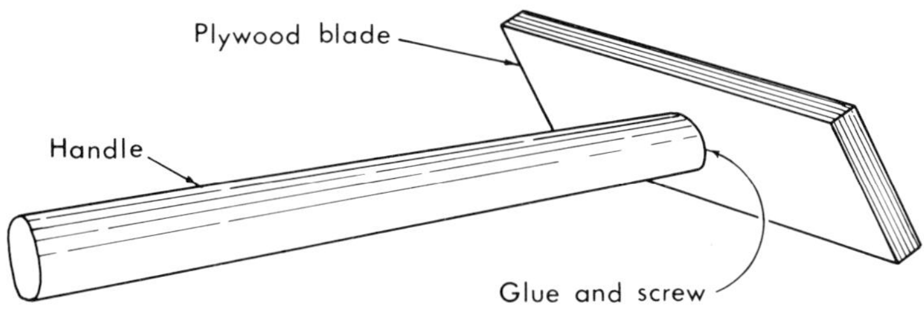

3. Home-made barnacle scraper.

Heeling

The vulnerable strip under the waterline can be scrubbed or painted by just heeling a boat slightly. While moored fore and aft, a light rope at the masthead usually suffices, but some ballast shifting may be needed to raise bow and stern alternately.

Short and long-handled scrubbers can be bought at marine stores. Have a few spares in case of breakage and in case keen visitors arrive at an active moment.

Barnacles are best removed by rapid strokes with a hand-made scraper (fig. 3) consisting of a small rectangle of $\frac{1}{2}$ in (12 mm) plywood glued and screwed to a short well-rounded wooden handle. Similar scrapers of steel are useful on badly fouled craft. Short pieces of batten and steel bar come in handy for awkward positions such as between rudder and sternpost and where bilge keels join the hull.

Boot-top

The boot-top paint strip of contrasting colour between antifouling paint and topsides receives punishment from lapping water, debris, weed and oil. After several vigorous cleanings, the boot-top may need some touch-up paint if smartness is desired.

Make sure all skin fittings are free from fouling and that zinc anodes for cathodic protection are clean and free from paint.

Cockpit

Any sunken well or cockpit below deck level makes a wonderful dirt trap. Bits of food and cigarette ash land along seat corners, in drainage pipes and gutters, and under teak gratings.

Although wiping with a swab rinsed in a

bucket of sea water keeps things sweet, whenever a fresh-water hose is brought aboard, the opportunity should be taken to give a self-draining cockpit a good douche, using full pressure down the gulley holes to free them of debris. However, vigorous use of the hose around the doors of under-seat lockers may dampen their contents – normal brightwork treatment with fresh water, sponge and chamois is kinder on these parts.

Many cockpits have lift-up sole panels giving access to engines or storage space. Waterways grooved into the bearers are piped overboard or into the bilges. This system may need cleaning every two weeks to prevent rain and spray from overflowing.

Fittings

The majority of boat fittings need no summer maintenance, particularly those of stainless steel, Tufnol, Delrin, nylon, or anodized alloy. These include such items as guardrail stanchions, tracks, blocks and winches. In general, moving parts consisting of plastics running on metal should not be oiled, whereas all metal to metal wearing surfaces need oil.

Chromium-plated fittings are popular on small powerboats and as portlights, door catches and hinges, they are also common on larger vessels. Whether the plating is over brass or die-cast fittings, deterioration of the surface can be rapid under sea air and salt spray conditions. To keep this at bay, treatment with car chrome cleaner followed by wax polish is advisable at least once a month.

Goosenecks, cleats, fairleads and similar fittings of silicon bronze and aluminium bronze are now in common use. Few owners bother to keep them highly polished (see Chapter 5) as they assume quite an attractive patina when weathered.

Older craft tend to have all fittings in brass, ordinary bronze, or gunmetal, including locks, hinges, ventilators and portlights. These metals turn green with verdigris when left alone and are often varnished to prevent oxydization. This does not produce a very attractive appearance, but keeping a lot of brass highly polished is a time-consuming job. A compromise is best. Polish the fittings which are always in view, such as binnacle and steering wheel; varnish, lacquer, or paint most of the others and leave fairleads, bollards, and chain pipes to tarnish.

Mortice locks, padlocks, and hinges on cabin doors and hatches which are exposed to the weather need a few drops of light oil mid-season. The same applies to other working parts, including screw-down ventilators. When checking a boat's rigging before, during, or after a passage, make sure that the locknuts on turnbuckles have not slackened, or that any seizing wires have not broken. Similarly, shackles, clevis pins, and wire rope strands deserve an occasional check. With the boat lying at a tidal wharf or staging, a close view of masthead fittings is often possible and a pair of binoculars may prove helpful.

If they run well, mast tracks and slides are best not oiled. When metal slides tend to stick, light oil helps, applied to the slides before hoisting sail. Thick oil or grease should be avoided, for although it stays in place longer, it attracts grit and is likely to stain sails.

Down Below

With several people on board, the saloon of a small cruiser soon gets dirty, particularly around the sole (floor) edges and under settee cushions or berth mattresses. Marina pollution regulations may prohibit the shaking of mats or carpets overboard. A low-voltage vacuum-cleaner makes a useful cruising yachtsman's birthday present!

Interior panelling, deckhead and doors rarely get any summer cleaning, but galley worktops and saloon tables need frequent swabbing. Food spillages around the galley are best cleared up immediately they occur to prevent drainage to inaccessible places with consequent foul smells.

If sea-water spray gets through an open porthole or hatch, the surfaces affected will stay damp in humid climates until swabbed down with fresh water. This applies particularly to a toilet compartment where the porthole or window is left open for ventilation. In any case, scrubbing around the *heads* at least three times a season is sure to be necessary. Although pan cleansing is beneficial, the use of caustic household chemicals can sometimes damage the pumps, piping, and valves, so get the right cleanser from your chandlery.

It pays to lay canvas sheeting over tanks and ballast stowed underneath cabin sole boards to stop crumbs, sugar, and similar particles from getting into the bilges. Cleaning is then easier and pests such as cockroaches are less likely to take up residence!

Dry bilges usually smell sweeter than moist ones, provided foreign matter is kept out. Some boats leak steadily and need regular pumping, others leak only when under way, while yet others remain almost permanently dry. Washing bilges with detergent should never be needed during a normal European or North American season but it does pay to remove any old food-can labels, bits of rag and other debris which could foul the bilge pump strum box (fig. 4) at a critical moment. Fortunately, beer cans (which yachtsmen like to stow in the cool bilges) do not normally have paper labels.

Stowage

The performance of most boats is readily affected by incorrect trim. The stowage of supplies for a long trip needs careful thought, and considerable experimentation is likely to be required with a new boat before arriving at the ideal distribution.

A small sailing boat trimmed down by the head often carries excessive weather helm, but

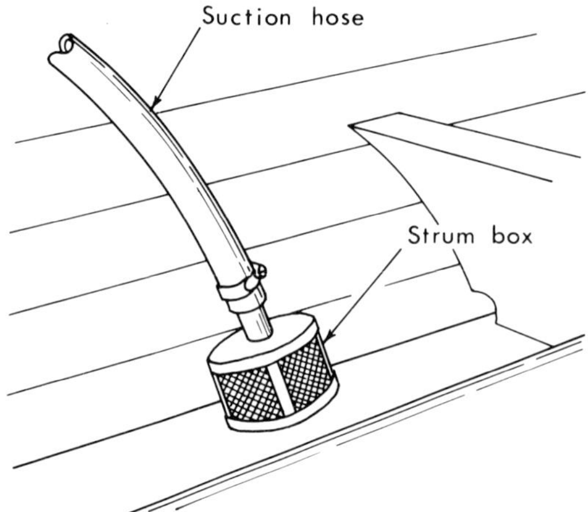

4. Position of strum box in pump well.

when day sailing, all personnel are likely to occupy the cockpit most of the time, producing the opposite effect. For extending cruising, heavy weights are generally best kept away from the bows.

Powerboats create great lift forward and the static trim is usually quite different from the cruising speed trim. Too much bow stowage may cause spray and green seas to come aboard. This condition also decreases the submersion of the propeller, giving reduced performance and perhaps dangerous handling qualities. On the contrary, small high-speed powerboats and inflatables can flip over if trimmed too low at the stern. Especially when you buy another boat, adjusting internal ballast is likely to become a summer care job.

Hazards

Safety precautions must always be in mind when in commission. The wise owner will shut off all seacocks when leaving his boat unmanned for a long time, to prevent siphoning back through the heads and to prevent trouble should a union leak, a pipe fracture, or a hose split. At the same time, battery circuits should be isolated except where alarm systems (with their own power supplies) must be kept alive.

Diesel fuel is not a big fire hazard and tank valves are usually left turned on to prevent the formation of airlocks. The reverse applies to petrol (gasoline) where isolation of tanks and regular checks for pipe leaks are essential. Bottled gas systems are notoriously dangerous and cylinder valves should be turned on only when an appliance is in use.

A gas detector in the bilges is a wonderful safety measure. Many good boats have been destroyed and injuries inflicted by unwarranted explosions. To check pipe joints and unions for leakage, smear all around with soapy water and watch for bubbles. Renew kinked or dented copper piping and eliminate flexible hoses except where essential to supply a gimballed stove or other movable appliance. Gas bottle lockers which are sealed (except for overboard vents) are safe, but make sure their

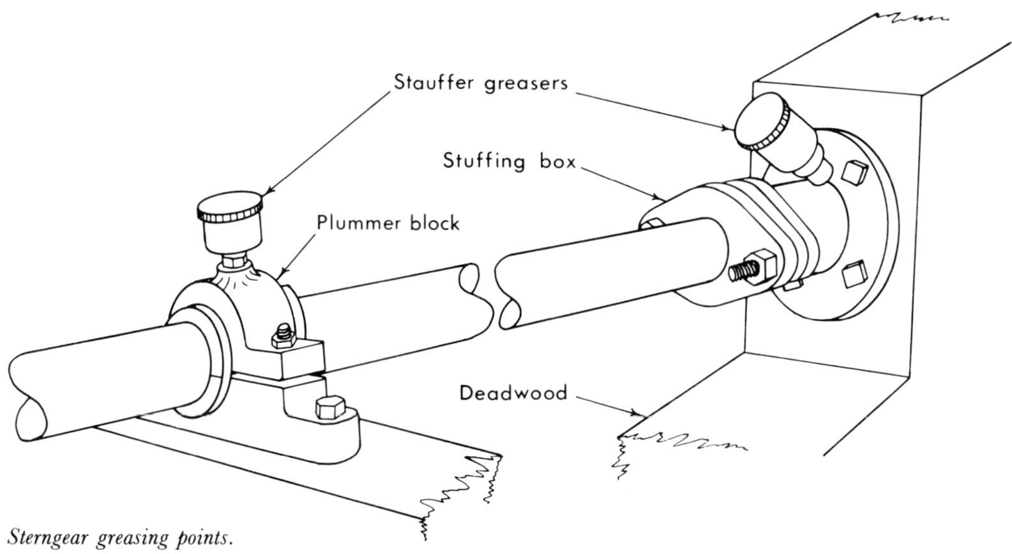

5. Sterngear greasing points.

scuppers are not bunged up with dirt.

Power

Lack of auxiliary or main engine care can prove a costly omission sometimes, while at other times it could lead to real danger. Basic servicing of a motor unit – according to the number of hours run – is only a small part of the summer care requirements. More breakdowns are caused by the failure of ancillary equipment such as fuel, cooling, and exhaust systems; electrics, instrumentation, and sterngear, than from troubles with actual engine units. Makers' handbooks do not often describe the maintenance of all components and it pays to make a check list applicable to your particular boat in order to keep a jump ahead of trouble.

When under way, feel sterntube bearings, plummer blocks, and thrust bearings occasionally for overheating and give grease cups a turn – see fig. 5. Make sure cooling water is flowing properly and that no fuel leaks are visible. On shutting down, dip tanks, tilt outdrive stern units and outboard motors clear of the water, and give stern glands a final greasing to prevent water dripping through.

The smell of a dirty engine can permeate

the whole accommodation with nauseating effects. Old engines are particularly susceptible to this problem and the wise owner will clean out drip-trays and wipe clean any parts moistened by oil or fuel whenever necessary.

On most modern cruising boats, battery failure leads to complete immobilization. A duplicate bank is a great safety precaution, but batteries stay healthiest when used and charged regularly, and extra care is necessary in this respect when there are two banks. Checking acid levels is always important, but especially for batteries which are of a considerable age.

Do not forget to oil the sheaves of steering cables and to check mid-season for slackness and broken strands. The lubricators on many power-operated pumps for water supply, bilge pumping, and engine cooling generally need a turn about every four hours of running.

Plate 5. Hatch cover protects varnish and prevents leakage.

Vandalism

Unless your boat is well guarded, make theft difficult by securing and covering outboard motors, battening down hatches from inside and ensuring the companion-way lock is of best quality and preferably duplicated. Stow loose deck gear such as the liferaft and lifebuoys down below if possible.

Make sure that all exposed instruments have strong covers and stow all removable instruments and other valuables out of sight in unlikely places or locked cabinets. Even if your mooring is in an isolated location, a burglar alarm can make any thief think twice before persisting.

Canvas covers over hatches, skylights and cockpit (plate 5) are a bother to rig, but they deter vandals and also protect varnish work. When a boat is left, closing all curtains and blinds stops prying eyes, keeps the interior cool and prevents the sun from bleaching polished or varnished joinery.

Ventilation

Even if some rain can get in, try to arrange for a circulation of air inside an unattended yacht. This may not be easy, as some small craft have no ventilators and propping a hatch open invites thieves. See also Chapter 10.

If a portlight is left open, make sure no bedding is beneath it. Leave cabin and toilet compartment doors open, but tie or wedge them to stop banging. Lifting the sole panels of cockpit and fo'c'sle assists circulation in the bilges and accommodation, but few owners bother to do this unless the craft will remain unused for several weeks.

Sailing Dinghies

Most racing dinghy enthusiasts maintain their boats carefully during the season, for they know that even slight defects make a lot of difference to race results. All sea water should be rinsed away at the earliest opportunity, not forgetting the sails.

Although plastic dinghies do not suffer when left salty, the resulting stickiness under humid conditions causes wind-blown dirt and sand particles to adhere. Furthermore, most such boats have seats and other parts of varnished wood which should be treated as described earlier under the heading *Brightwork*.

Provided there are no leaks, removing the bungs from the built-in buoyancy compartments of a fibreglass dinghy is not necessary, but ventilating in this manner is wise on any wooden dinghy which has an overall cover to keep the rain out. Without a cover, remember to leave hull drains or self-bailers open, and chock the boat so that water will drain out of them.

Rowing Dinghies

Whether inflatable, folding, or rigid, a yacht's tender normally receives little in-season care, except for an occasional wash down. After several landings on sandy or muddy shores, bottom boards and bilges get grimy. There is no quicker treatment than to take her to a convenient beach or hard, remove the bottom boards, get someone to help you up-end her with transom downwards and bow in the air (plate 6), then hurl several bucketfuls of water into her. Any residual dirt can be swabbed from around the transom before letting her down.

Wash the bottom boards separately before replacing them and, if you have time, scrub the all-round fender (or the outside of an inflatable boat) to remove grit which could scratch the parent vessel. Cleaning the topsides could be done at some future time, but the outside of the transom is certain to need dowsing. If the bottom needs cleaning, turn the dinghy upside down – preferably before any of the above tasks are started.

Even in dry, calm weather, dinghy seats become wet from splashing oars and dripping

Plate 6. The simplest way to clean out a tender.

lines at the landing steps; always keep a cheap sponge or old towel handy.

A dinghy painter can chafe through unexpectedly, so check regularly and renew when doubtful. Many a tender has been lost while towing and a second painter is well worth rigging when passage making.

Inflatables

Inflatable boats generally create few problems provided no leaks develop or bonding failures occur to fittings such as painter eye, inflator pipes, oar crutches, or outboard mountings.

Leaks may be detected by submersion in water, piece by piece, when partially deflated. Most makers of such craft supply temporary repair kits with full instructions. For repairs of the pin-hole type, a patch can remain sound almost indefinitely without risk, but if the boat has to be returned to the factory with major damage, you may be able to keep going by hiring a similar boat.

As with their rigid brethren, rubber dinghies need an occasional wash – preferably using fresh water. Removing grit from inside is sometimes impossible even when up-ended and the quickest procedure is to hold her upside down in mid-air while playing a hose into her. When no hose is available, a possible alternative is to use a small portable bilge pump with rubber pipe outlet and operate this vigorously with the intake end submerged in a pail of water.

One of the merits of a rubber dinghy is the ability to deflate her for convenient stowage on board a cruising boat. However, re-inflation is such a difficult process using standard air pump or bellows that many skippers treat their inflatable as a rigid tender – towed behind or stowed flat on deck. Sophisticated methods for rapid inflation are possible, using skuba-diving cylinders, or a compressor powered by electricity or the boat's engine. Anyone who has attempted to inflate a rubber dinghy by mouth will think it prudent to have a spare set of foot bellows to guard against sudden failure of a one and only set!

Multihulls

Cruising catamarans and trimarans often have vastly greater deck and hull surface areas to maintain than monohulls of comparable length. However, due to the narrowness of the hulls, internal maintenance is not much greater, provided there is access to all parts and sufficient space for a person to work.

Keeping multis to their designed weight is imperative for safety and good performance, so extra in-season care is necessary with regard to trim and the stowage of stores and gear. Soakage of water by wooden hulls increases weight greatly and is a critical factor with racing catamarans. Such hulls must be kept dead tight – polyurethane coated inside and possibly fibreglass sheathed outside. If water gets in, one may have to open sealed compartments after each race and dry out the hulls with a warm air electric blower.

Due to their stiffness, multihulls suffer high rigging stresses, so frequent inspections for defects are prudent. The beams joining hulls together are all-important. Their connections need checking regularly for slack bolts and signs of strain – especially when the beams are of the adjustable or removable type.

Job List

Most of us get too little time for boating. One of the most useful aids in preventing wastage of those precious moments is a small notebook – much more reliable as a memory-jogger than the backs of old envelopes.

To do the job properly, divide the notebook into three parts – front pages for a list of items to bring aboard next trip (such as stores, books, and tools); middle pages for in-season jobs to be done; back pages for the not-so-urgent jobs which are normally left until winter. Crossing out each item when dealt with is satisfying, and tearing out completed pages helps towards rapid reference.

Rectifying defects is part of safety afloat, but the memory plays tricks and even essential jobs may be overlooked if not jotted down in the notebook. Whipping a rope's end, replacing a broken hinge, stitching a sail, or swinging the compass, can be enjoyable tasks on a quiet day at anchor, but they could all lead to trouble if delayed too long.

three

Autumn Care

Some sailing goes on where the climate is kind enough to permit cruising and racing almost every month of the year. For all other sailors, the off-season of cold weather, heavy rain, monsoons, or hurricanes lasts from two to six months.

Storage

Boats are best stored under cover during the off-season, but things do not necessarily stop there. The wise owner will carry out a satisfying overhaul to ensure that the need for summer care is minimized and all defects listed during the season are rectified.

Ideal storage space is expensive nowadays and those who change boats frequently, often expend minimal time and money on maintenance, leaving their craft afloat continuously. The majority of owners with boats too large for trailing manage with a mud-berth sheltered from wind and tide, or perhaps rent space on a quayside or at the back of a sailing club or boatyard.

In temperate or cold climates, to keep a craft in first-class condition, especially when under about 50 ft (15 m) in length, she is best hauled out during the winter. Unfortunately, due to the lack of facilities and the high cost involved, thousands of craft are doomed to stay afloat perpetually. When living aboard, this does little harm, as lines and fenders can be checked regularly, ice or debris warded off, and most parts below decks kept warm, dry and well ventilated.

Metal parts below water are often subject to galvanic corrosion – see Chapter 4. Prolonged immersion accelerates this, though the rate of damage is slower in winter when the water is cold. Mooring an alloy or zinc-sprayed steel boat near to one with copper sheathing or copper antifouling paint could lead to serious galvanic action.

Much trouble has been experienced with certain fibreglass boats left afloat continuously. The skin absorbs large quantities of water and the gel coat becomes covered with hundreds of blisters. This can lead to costly repair work, see Chapter 9.

At some river moorings, floating ice can damage hulls of all types. Weighted boards suspended at the bows help to divert ice floes – even better if they are fitted all along the waterline as well.

Trailing

A popular and cheap method when laying-up craft under about 36 ft (11 m) in length is to move them by road to one's own backyard. A special trailer can be bought for the purpose and the boat left on it during the winter. Alternatively, a trailer can be hired for the appointed day, jacking the boat on to chocks to release the trailer – see plate 7.

Lifting from the water is simple using a boatyard crane. With a powerful towing vehicle, it may be possible to slip the trailer

Plate 7. Jacking up a boat to permit withdrawal of trailer.

under the floating boat on a launching ramp and draw both up to level ground. Hosing down is not always sufficient to clear the salt water from the bearings and brakes, so stripping these down soon afterwards is usually advisable. Greasing bearings fully beforehand is a good precaution. Remember that brakes rarely function properly until fully dried out.

Most sailing dinghies have their own trailer and launching trolley. The latter should have wheel bearings which are not affected by water. The trolley is often used also to support the boat between races when stored in dinghy pen or boat park. It should be possible to intermesh the two carts, enabling the boat to be transferred from trolley to road trailer (or vice versa) with a minimum need for direct lifting.

Similar arrangements can be used for small cruisers or speedboats and special trailers with winches, jacks and rollers attached are available to permit launching and recovery single-handed. Car roof racks with built-in rollers enable light dinghies to be transported without trailers.

Stripping Gear

Boatyards and marinas often provide cabins and racks for the storage of removable gear and equipment, such as vane steering gear, outboard motors, anchors, inflatables, sails, masts, mattresses, fuel cans, chains, ropes, and fenders. Instruments, charts, books, blankets, mats, food, and clothing are normally taken home. Some owners take all gear home except mast, spars, and dinghy.

Moving the small items is quite a big job and it pays to contribute to this every time the boat is visited towards laying-up time. Tools and stoves are often left on board for winter use. If the anchor cable is to be left on board, range it all on deck or on the ground for examination and until any drying out, cleaning, or painting of the cable locker has been completed.

Wilful damage can occur on laid-up craft, though rare when all movable equipment has been taken away. However, do not be too complacent, for it has been known for propellers to be sawn off; also deck winches and other fittings have disappeared before now. When a boat is ashore, a speedometer impeller could be damaged even when withdrawn into its housing. It pays to drive plugs into skin fitting openings to stop small boys from stuffing mud and stones into them!

Hauling Out

Whether by crane, travelling hoist, or inclined slipway, amateurs usually leave hauling out to the professionals and restrict their own activities to the use of trailers. However, should one have to organize the whole operation in remote parts, there are no great problems in laying down a pair of greased timber *standing ways* (nailed to cross ties – see plate 8) plus a timber cradle with *sliding ways* of correct width. Arrangements can be made for a hand winch, powerful truck, or farm tractor, to be available at the top of the ramp.

Note that a timber cradle needs ballasting with scrap iron or rock to submerge it while the boat is floated into position on the next high tide for hauling out. Unless the cradle is to be left beneath the boat until spring, it must be fitted together with bolts to facilitate dismantling while the boat is jacked up on chocks.

Frequently-used boatyard standing ways are often of steel rail or angle iron, the cradle having four steel flanged wheels to match.

Plate 8. Greased standing ways on the foreshore.

Striking Over

The biggest boat can be moved any distance – even around corners – by shifting greased ways where required and re-locating the winch, or fitting snatch blocks to the hawser.

With the cradle dismantled and the boat on chocks, athwartships baulks can go under the keel to allow striking over sideways into position between other craft. Slowly but surely, crowbars (levers) will enable small craft to be struck over any distance, but hydraulic jacks angled between ground and keel must be used for big vessels.

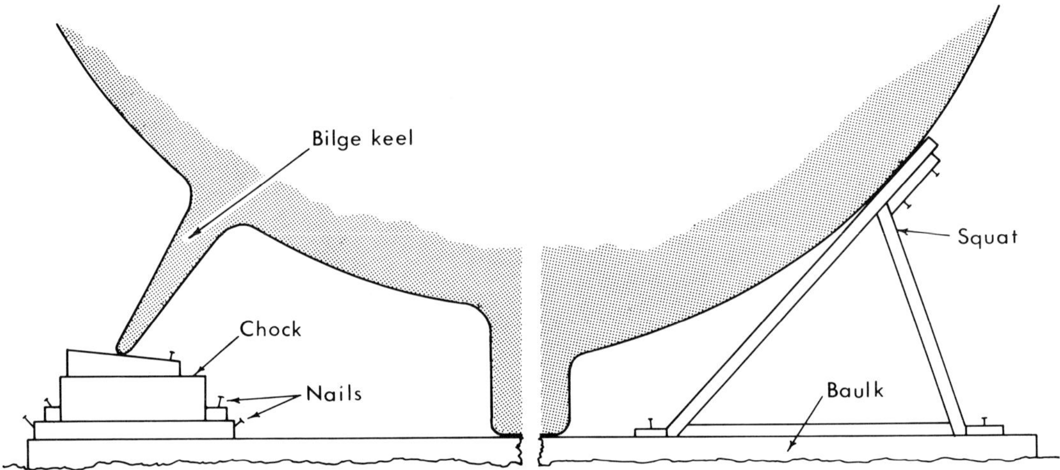

6. Two types of hull support squats.

Legs, or supports under the bilges, are needed to keep her dead upright, adjusted and freed occasionally as the movement continues. After slight jacking to remove the baulks and replacing them with chocks, *squats* must be firmly nailed together under the bilges (fig. 6) unless the boat is equipped with legs (fig. 7) bolted to the topsides. Note that legs can slew and allow the boat to collapse unless fore-and-aft guy ropes are rigged. Legs get in the way of topside painting while squats obstruct work under the bilges. Therefore, a change over from one method of support to the other is often desirable.

Slipping Preparation

Before hauling a boat out, masts should be unstepped or lowered in their tabernacles. If the boat is old or delicate, her weight should be minimized by draining all tanks and removing inside ballast.

Once clear of the water (or as soon after slipping as possible) the bottom should be scraped and scrubbed clean as described in Chapter 2. If allowed to dry hard, weeds and barnacles take four times as long to remove. Furthermore, if a sander is used on antifouling paint, the flying powder proves injurious to skin, eyes, and lungs.

A sea-going boat needs hosing down all over with fresh water while on the slipway, to remove the hygroscopic salt which will prevent drying out, attract dust, and make preparation for painting more difficult.

Rigging

Before unstepping the mast, all standing and running rigging must be disconnected from the boat's deck and frapped to the mast – not forgetting any electrical cables to navigation lights, spreader lights, or wind speed and direction indicator. The forestay and a pair of shrouds should be replaced temporarily to prevent the mast from swaying while removing the coat and wedges which secure the mast at the deck partners, or to prepare tabernacle bolts for withdrawal.

Unstepping the Mast

A mast stepped through the deck needs to be lifted bodily upwards by crane or sheer-legs to free it. Suspension must be from a point above half the height. The best way to achieve this is to use a rope strop rigged as in fig. 18. A marline hitch is urged up the mast (using a boathook or spinnaker pole) while the rope tail is flicked and slack is taken up on the crane hoist. Then, the bottom end of the sling

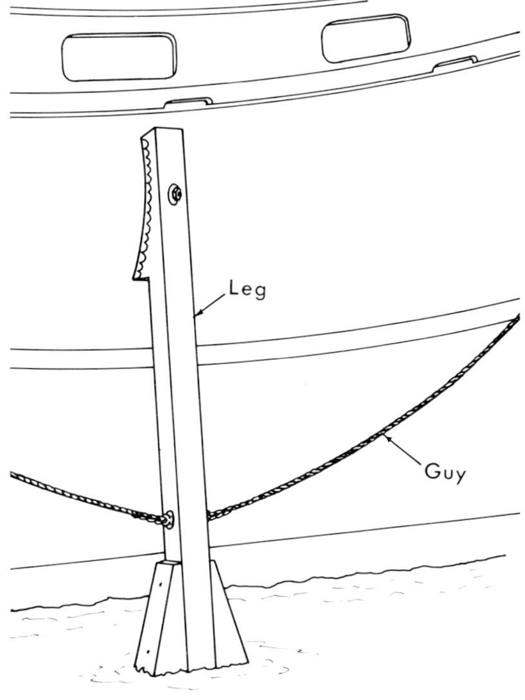

7. Ensure that legs are safely guyed.

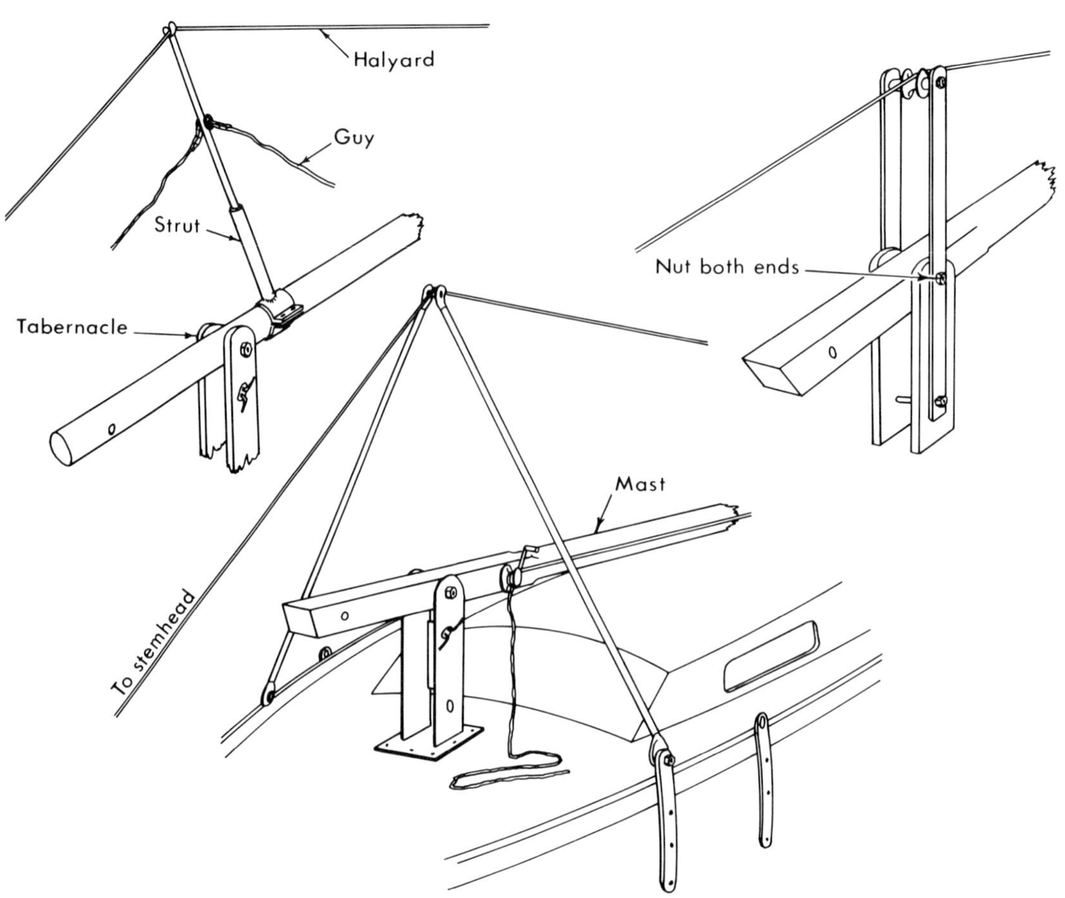

8. Three types of struts for raising and lowering tabernacled masts.

is made fast beneath the halyard cleats or gooseneck.

If sheer-legs of sufficient length can be borrowed, lash their feet to deck fittings for safety's sake. Swing the mast horizontal then manhandle it down on deck, or across planks on to a quayside which is approximately at deck level.

A mast stepped into a socket on deck only needs lifting a few inches to free it. Lowering a tabernacle mast is even easier for, having removed the bottom bolt, the top bolt may be used as a pivot. Hinging the mast down on deck may be accomplished by means of a jib halyard attached to the stemhead. However, to give the halyard a safe angle as the mast comes down, use a special strut (or A-frame) rigged as shown in fig. 8. If a single strut is used, great care must be taken to ensure that it has adequate staying to prevent sideways collapse under load. Such a strut may be stowed on board to facilitate lowering the mast should the vessel need to pass under a bridge.

When a crane is available, a tabernacle mast is best lifted vertically and dumped where required.

Undressing

Nowadays, alloy masts are frequently stowed for the winter with all rigging and fittings (which are of unperishable materials) left in place. However, to minimize theft and to enable all parts to be freed from dirt and salt, the owner is advised to *undress* his mast, coiling and labelling each stay, line, and fitting. These should be taken home, scrubbed in soapy water, rinsed and hung to dry. Where halyards are housed inside a hollow mast, pull in temporary cords to simplify re-rigging in the spring.

A wooden mast will have to be undressed anyway, being certain to need partial or entire re-varnishing. Make sure all spars are well supported and straight when stored on racks or slung under a shed roof. The short masts of motor cruisers are frequently left in position unless the shed has insufficient headroom. If the boat is stored in the open, all tophamper may have to be stripped off (perhaps including radar scanner and davits) to enable her to be cocooned with tarpaulin sheets.

Mudberths

A cheap and safe (though often inconvenient) method of laying-up a seagoing craft is to motor or tow her on top of a high tide into a shallow creek preferably near a boatyard, house, or farm, where someone can kept an eye on things.

Considerable preparation is necessary, especially on the first occasion that a certain spot is used. A *wallow* should be dug, by hand shovelling, to receive the keel and perhaps anchorages fixed for mooring lines. For the latter, stout posts of wood or steel tube work

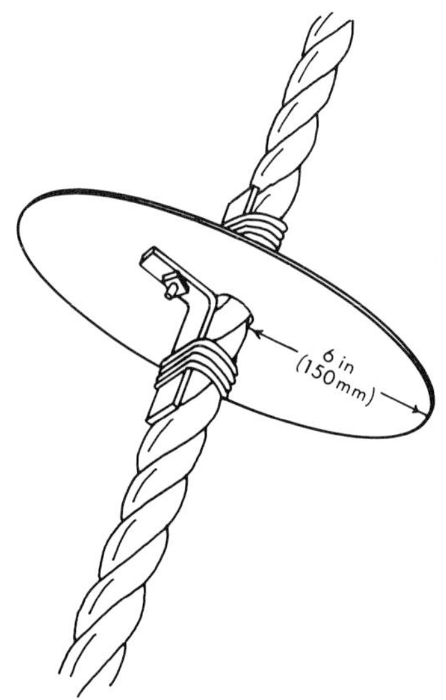

9. Tinplate disc to discourage rats from getting on board

well in firm ground, but in softer conditions better security is provided by digging in heavy old boat anchors or *deadmen* – baulks of timber buried in trenches at right angles to the line of pull, having a wire rope strop around the middle which is led above ground some distance away before backfilling.

Normally, two lines should radiate from the bow and two from the stern to keep a boat on station, with just sufficient play to allow her to rise on a very high tide or river flood. Slack wire hawsers should augment the relatively taut synthetic lines in case chafe parts a rope one stormy night. Discs of tin-plate fitted to ropes as shown in fig. 9 will prevent vermin from seeking shelter on board.

If no extra high tide materializes at laying-up or fitting-out time, it may be necessary to dig an extension to the wallow or use a powerful winch to get the boat in and out of her mudberth. Note that slight pitting of the metal surface has been reported in some places when polished light alloy boats are stored in mudberths or lie ashore with the plating in contact with dry soil.

Covers

Whether dinghy or decked boat, an over-all cover is essential in most climates if she is stored in the open. Without this, she will remain damp below, her brightwork and decks will suffer, frost may cause cracks and leaks, while vandalism is more likely to occur. Solid ice in the bilges should be prevented at all costs – complete dryness is the best safeguard.

Hulls of plastics, metal, and ferrocement may be left high and dry on shore at any time of year without deteriorating appreciably. Drying out once a year is the healthiest possible treatment for any type of wooden boat provided this occurs only in winter. In Britain, Canada, and the northern parts of the U.S.A., such craft must be returned to the water not later than the end of April if damaging con-

Plate 9. Ideal framework for a winter cover.

traction of planking and backbone is to be avoided.

A custom-made cover is standard equipment for the average sailing dinghy. For cruisers, most owners get by with standard waterproof sheets, well tied down over a timber framework.

New materials appear from time to time, but four main types are in common use for boat tarpaulins – heavy duty translucent polythene, thin nylon netting imbedded in pliable PVC, proofed cotton canvas, and PVC-coated nylon cloth.

The framework usually consists of a central ridge pole (*strongback*) supported by vertical posts lashed to tabernacle, deckhouse, companion hatch, windlass, and through-deck openings, plus raking struts from ridge pole to deck edge or guardrails – see plate 9.

Any slackness of the tarpaulins allows the wind to play havoc. Therefore, eyelets need to be numerous and strong with lashings passed right under the keel, or made fast to a guest warp right around the hull, with a few underkeel lashings to prevent this from riding up.

Air Currents

Both ridge pole and tarpaulins must overhang at bow and stern to exclude rain and to incorporate ventilation apertures. An old polythene bucket with the bottom cut out makes a good air duct when the canvas is tied around it.

Every hatch, door, locker, removable sole (floor) board, and bunk bottom should be opened up for maximum ventilation, except when someone is working on board. Tarpaulins cannot be folded back while work progresses in wet weather, so panels of translucent plastics (or complete covers of similar material) are then of enormous benefit. Large press-in moulded eyelets can be bought, enabling ropes to be attached to the edges of standard polythene sheeting.

When electricity from the shore is available, a small heater left on continuously below decks is useful in damp conditions, particularly when a craft is sheeted over while laid-up afloat and perhaps used occasionally for sleeping aboard.

Dinghy Pens

A dinghy racing enthusiast living in an apartment may have to lay up his (or her) craft in the club dinghy pen. With no cover, such a boat could be stored upside down resting on four old car tyres. If a cover is used, precautions must be taken to prevent rain trapping. If necessary, fabricate a simple ridge pole assembly as described previously.

The mast must always be unstepped to prevent blowing over, unless stayed to anchorages well away from the boat. An upright boat should be taken off its launching trolley and supported on car tyres under the bilges, with two or three wooden chocks under the keel. All buoyancy compartment hatches must be left open, also self-bailers and drain plugs. Inflatable buoyancy bags, loose gear, running rigging, and sails are best stored at home.

A rowing dinghy stores successfully standing upright on her transom with the open side towards a high close boarded fence, double lashed to big galvanized eyes screwed into the fence posts. An inflatable rubber dinghy needs careful autumn examination for damage in case she needs returning to the makers.

Inside Ballast

Most motor and sailing cruisers carry varying amounts of inside trimming ballast, which is usually left in position when laying up, see Chapter 4.

If bilges are to be cleaned and painted, it pays to remove all ballast and bilge tanks on hauling out. Then the bilges can be washed with a hose and nozzle, shifting dirt far more easily than when everything has had time to dry out. If there is a drainage plug aft with nut and washer inside, remove this to eliminate constant pumping. Alternatively, *syphon* the water away through a garden hose, leading over the deck and down to ground level. To start this, fill the hose with water. Hold your thumb over one end and arrange for a helper to do the same the other end. Install the hose, then, at a given signal, release both ends in unison. This is a useful way to drain a dinghy which has become filled with rain water while on shore.

A portable hand bilge pump comes in handy to remove the last dregs of bilge water (or to shift the water past a watertight bulkhead) when the hull is chocked up at an unnatural angle.

Before removing ballast, the pigs of iron or lead should be given painted numbers and a small plan prepared to indicate their positions. This is especially important for the older type of sailboat which has part of her bilges almost completely packed with ballast.

Ballast is usually kept away from the planking by means of fore-and-aft battens laid across the frames or ribs. In some seaboats, the ballast is battened over to prevent it shifting in the event of a knock-down in severe weather.

Anchor Lines

The smallest boats often use nylon anchor cable with a short length of chain next to the anchor. Craft frequenting very deep anchorages use stainless-steel flexible wire rope stored on a windlass drum, also with chain at the anchor end. Most other boats and fishing craft have tested galvanized chain all the way, the inboard end shackled or lashed to an eyebolt in the chain locker to prevent accidental loss of both anchor and cable.

Chain is usually unshipped during laying-up (see Chapter 4), but keep it under lock and key to avoid theft!

Batteries

The massive batteries fitted to some big powerboats are often left on board and charged regularly using the ship's charging set, a portable generator, or through a charging panel actuated by a shore supply. The batteries for smaller craft are better taken ashore, cleaned down, trickle charged at home or at the boatyard and topped up when necessary. Alkaline batteries are not so delicate as the common lead-acid type. Remove all dry batteries from equipment to prevent corrosion damage.

Engines

If engines are to be overhauled, removing by crane before slipping proves easier than using sheer-legs in a restricted shed or yard. If a boat is laid-up outside, the electrical components should be taken off and stored at home. This applies particularly to magnetos but, to be on the safe side, do likewise with starters, dynamos, alternators, windmill generators, distributors, coils, spark plugs, switch panels, air-conditioning units, and electric pumps. Outboard motors and small charging sets are also best stored at home if not needed at the boat. Various aerosol sprays are available for inhibiting rust on engines and preventing corrosion and verdigris on electrical equipment.

Sails

Dinghy sails are fairly easy to wash in a bathtub to get rid of salt and dirt before drying and storing. Unless big cruiser or ocean racer sails can be dunked in a swimming pool for a day and then scrubbed on a lawn before drying, it will be wise to return them to a suitable sailmaker for cleaning. Cotton and flax sails are best draped loosely to allow air circulation when stored, making sure there are no rats or mice about. To save double handling, the sailmaker might as well check the need for any repairs and undertake these well before the spring rush. He may have facilities for sail storage – a most useful service for certain owners.

Sails of Terylene or Dacron are notorious for chafed stitching due to the hardness of the cloth. Any such sails not sent away for general repair should be examined minutely for faults. Full details of all sail problems are given in the companion volume *The Care and Repair of Sails*.

The Liferaft

Most makers of emergency inflatable liferafts recommend owners to return them to the nearest service bay at laying-up time for examination and for re-packing into the container from which they spring when activated. In any case, if the raft has been inflated in an emergency (or in a rehearsal) it should be sent away for recharging and correct packing.

four
Winter Care

While laying-up, no doubt the old summer job list (see Chapter 1) will have lengthened, as faults to rigging, unloaded gear, exterior damage, and loose paint are seen from a different angle.

Avoiding the Snow

At this time of year, northerners may envy enthusiasts from the south (or vice versa for the southern hemisphere) who are still able to sail and to do a bit of varnishing under ideal conditions. There may be compensations, however, for having a winter overhaul means that a boat is kept in better condition and the hobby is not likely to get monotonous!

The owner with an old boat and too little money often looks forward to the out-of-commission period as a time when the vessel's rejuvenation can be taken a stage further – a complete engine overhaul might be the next job – ensuring an enjoyable season the following summer.

Examination

A recently acquired yacht with no survey report needs a thorough examination the first winter to ensure that any defects are dealt with well before the mad spring rush starts. The long-term owner is not being unduly cautious if he engages the services of a professional surveyor every five years. Even a modern fibreglass hull can be suspect, for unwitnessed collisions are occurring all the time at moorings, while any part of the moulding could have an intrinsic fault which only shows up by blistering or crazing several years after construction.

Watch a planked hull drying off just after hauling out – any porous seams will remain damp long after the others have dried. Mark the faulty seams with chalk and re-caulk them during the winter.

Tap along each plank at close intervals with a small hammer, also along keel and stem. A solid resonant sound means all is well; a dull hollow sound suggests rot. Mark suspicious parts with chalk, then try piercing the surface with a thin bradawl. Easy entry will confirm a defect, but even if the surface seems as hard as surrounding parts, try to locate the corresponding spot inside the boat and test again by prodding there.

A light tapping test will divulge lamination faults in fibreglass, but considerable experience is necessary for complete success, as skin thicknesses often vary over a hull, while built-in frames, brackets, and timber inserts upset the tone.

Galvanic Action

The only way that an aluminium hull of the correct marine alloy can corrode is by galvanic (electrolytic) action, but this process can have far-reaching effects on the metal com-

Plate 10. The results of galvanic corrosion.

ponents of nearly all craft. Careful winter examination for signs of pitting is essential. How to rectify such damage is described in Chapter 9, but when there is trouble a system of cathodic protection is called for – see Chapter 8.

Most people know how a simple battery is constructed: a positive plate (cathode) and a negative plate (anode) surrounded by a chemical liquid or paste. Any dissimilar metals in salty or polluted water form a battery, one metal (the anode or less 'noble' of the two) being eaten away as a current is generated. This can happen to the underwater parts of boats, inside or outside.

The speed of destruction depends on the nearness and size of the two parts, the impurity and temperature of the water, and the types of metal. The relative activity of some common marine metals is depicted in the following table.

Base (Anodic) End	Zinc
	Galvanized steel
	Cadmium plate
	Aluminium
	Light alloy
	Mild steel
	Cast iron
	Stainless-steel (active)
	Babbitt metal
	Lead
	Manganese bronze
	Brass
	Copper
	Silicon bronze
	Gunmetal
	Nickel
	Monel
	Stainless-steel (passive)
	Chromium plate
Noble (Cathodic) End	Graphite

With any two metals in sea water, the higher one on the list gets damaged while the lower one is protected. The nearer together they appear on the list, the slower the action. It should be noted that some types of stainless steel are activated by contact with other noble metals, suddenly becoming anodic to any of the metals below cast iron on the list. A curious contact corrosion may also occur where stainless steel passes through rubber bearings or fittings of plastics.

Efficient under-water paint goes a long way towards eliminating galvanic action, but any holidays (missed paint) or abrasion can lead to greatly increased activity on the small area exposed.

Propellers

Galvanic pitting of bronze propeller blades (plate 10) is aggravated by the phenomenon called *cavitation*. This is created by a propeller spinning too quickly to keep in contact with the water.

A badly pitted propeller, or one with buckled, serrated, or chipped blades, should be sent away for rebuilding; any good boatyard or chandlery store will arrange this. To free a propeller from its shaft taper and keyway, remove the cotter pin and retaining nut, then apply a sharp heavy hammer blow via a hunk of hardwood resting against the propeller boss. If obstinate, you may have to jack it off with a proper pulling tool, augmented by hammer blows. Without a puller, two people hammering in unison to port and starboard is best to prevent bending a thin shaft. But do not let the propeller fly off – if insufficient help is available, use a hanging cord tied around one blade or leave the nut on loosely.

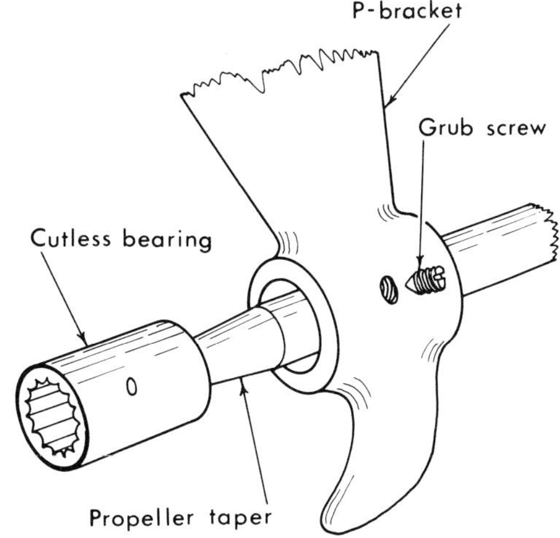

10. Cutless bearing withdrawn from a P-bracket.

To check a propeller shaft for wear in the outboard bearing, unbolt the engine flange coupling and push the shaft outwards a little. If wear is observed, draw the shaft right out, removing the rudder if necessary for clearance. Getting the shaft built-up and skimmed to original size normally proves much cheaper than having an entirely new shaft machined.

Cutless rubber bearings (fig. 10) tend to

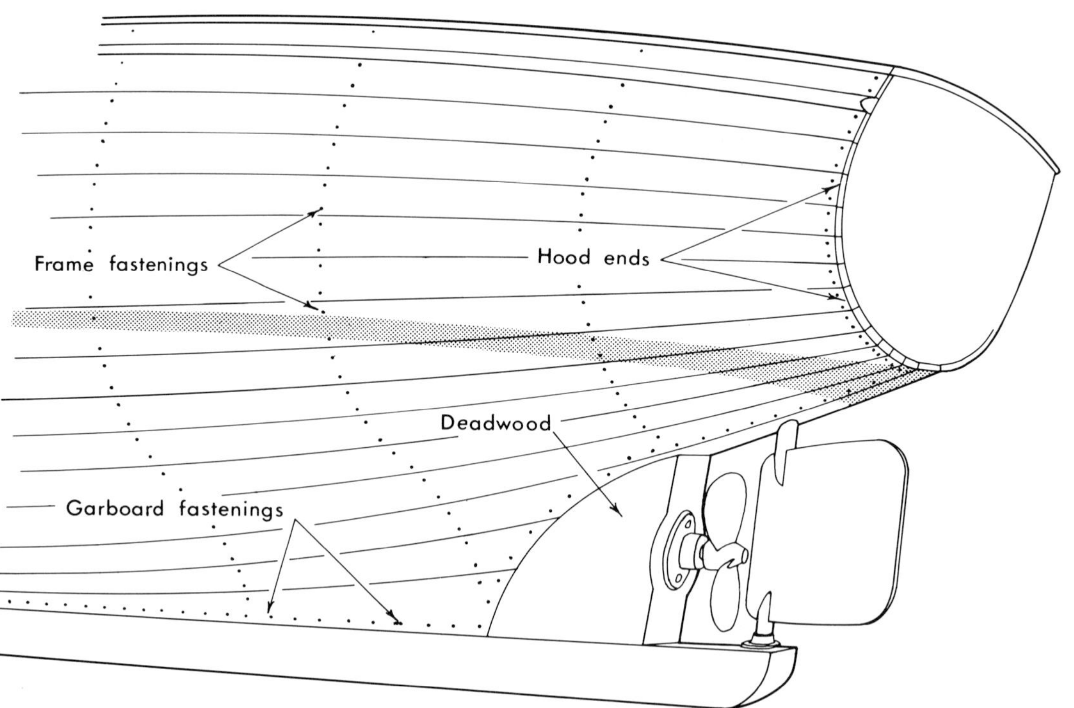

11. Some important fastenings.

wear rapidly in silty or sandy water. They are easy to push out of a P-bracket, but with a sterntube it generally entails withdrawing the shaft.

Fastenings

When you can pick out the locations of underwater plank fastenings on a painted timber hull in winter, trouble is looming. This occurs most frequently where copper fastenings were used to secure galvanized steel floor knees across the keel, running up three or four strakes of planking. Galvanic action softens the timber around each fastening and eventually leakage starts – see Chapter 8.

Amongst other suspect fastenings could be the hood-end screws or nails where each plank terminates at the ends of the boat, also the keel rabbet (rebate) fastenings along the bottom edge of the garboard (lowest) planks – see fig. 11. When such fastenings have *started* (loosened) – probably due to a collision – the first indication of trouble is most likely to be plank edges standing up proud of the adjacent surfaces. Mark with chalk for early attention.

Bolts

Although it may not tell the whole story, a visual check on the bolts, screws, and clenches holding rudder gudgeons and pintles, skeg, bobstay fitting, bilge keels, chain plates, sterntube bearing and skin fittings, sometimes reveals looseness or the start of corrosion. As well as old age, trouble may be due to the use of faulty or incompatible materials, accidents, timber shrinkage, or the over-enthusiastic use of power sanders. On old boats, it pays to renew a few of these fastenings periodically to be certain of their condition. Craft built of materials other than timber are by no means immune from such defects.

Bolts securing massive iron, lead, or concrete ballast keels are often a worry on vessels more than about twelve years old. It does not take as long as most owners think to drive out a couple of keel bolts at random, then replace or renew them. However, one can usually find (by scanning the boating magazine advertisements) the whereabouts of firms equipped to X-ray keel bolts without needing to disturb them.

The time to worry has arrived when rust stains come from the joint between metal and wood keels, or when the once-galvanized keel bolt nuts inside the boat have become misshapen masses of rust which can no longer be turned; see Chapter 8 for further information.

The deep bilge keels, fins, and skegs fitted to some sailing craft (and the shallow bilge keels used to lessen the rolling of some powerboats) have bolts through the hull which sometimes deteriorate more rapidly than the main keel bolts – remember to test them also.

Damage

After hauling out, an external examination is bound to reveal unexpected damage which needs adding to the job list. Metal boats often show scars in the form of dents perhaps made by the sharp stems of racing dinghies trying to tack too close. You may decide to do nothing about these. A similar blow on plastics usually leaves star-shaped cracking; ferrocement might chip locally, while the surface of thick plywood or timber planking could get gouged in varying degrees.

Sheer beadings and rubbing strakes receive severe punishment sometimes alongside quays and rusty steel piling: anchor flukes score topsides, chains catch around the stem, barnacle encrusted mooring buoys act like magnets – some highlights of summer return to mind when the winter repair jobs are listed! The aim is to probe into all defects to find out which need structural repairs and which can be fixed with stopping during annual painting.

Inside

Other than basic housework, newish boats need little interior winter attention. Some of the modern materials used for linings, worktops, deckhead, bulkheads, and doors never need painting. Older craft often need a programme of interior work lasting several winters. This ensures that the smell of paint and varnish is largely dispersed before the next season. If just one compartment is painted each year, this will ensure that the intensity of smell (which proves highly objectionable to certain people) is minimized.

With all the crevices, obstacles, knees, removable panels, lockers, doors, shelves, sole boards and bunks, even a small cabin takes a great many hours to paint properly. Therefore, avoid the temptation to tackle too much at one go.

Including varnish, one cabin could require the use of four or five different types and shades of coating.

To limit condensation, hulls of metal, plastics or ferrocement which have not been lined throughout with bonded-on insulation materials may need special anti-condensation paint (containing cork granules) against the hull skin and framing. Alternatively, this may be a good time to get a thick layer of foam plastics sprayed on, or to glue light-weight insulating tiles over the offending areas.

Washing Down

The day when galley, deckhead and bulkhead painting becomes necessary can sometimes be deferred by several years if a bath-tub cleaner is used to wash the surfaces down. Food lockers generally need a good scrub each winter, while most shelves, drawers, and hanging lockers benefit by wiping out with a damp swab followed by drying with an old towel.

The toilet compartment needs extra treatment to keep it sweet. The lower paintwork (and certainly the sole boards or their covering) will need thorough scrubbing every winter.

Dirty Jobs

As interior cleaning is best started from the deckhead downwards to get rid of the drips,

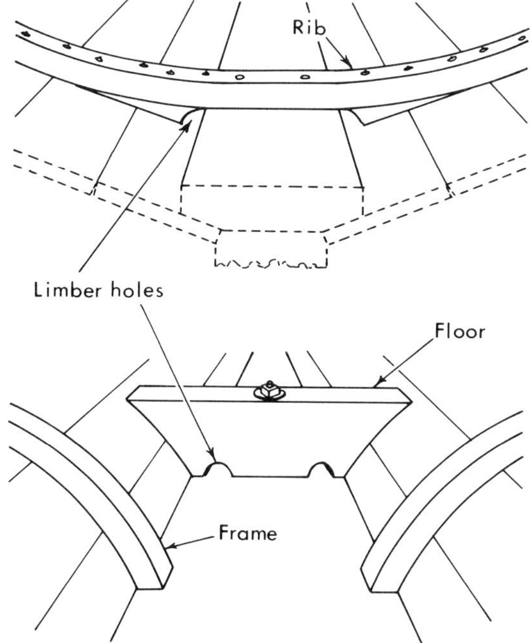

12. Common limber hole positions.

bilge cleaning should come last. The routine for winter bilge cleaning varies greatly, according to hull material, whether there is any inside ballast and whether any major spillage of food, oil or fuel occurred during the season.

Having hosed down the bilges at laying-up time (see Chapter 3), try to avoid repeating the process because complete drying out might take several weeks. From amidships forward bilges normally stay relatively dry and clean. Abaft that, oily bilge water may have left its scum and a good scrub with a proprietary bilge-cleaning detergent, followed by hose jetting, will be essential; as mentioned previously, this is best done in the autumn. The bilges of timber and steel craft are usually painted and this should last three or four years.

Pigs of iron ballast must be coated to prevent powder and flakes of rust from dropping into the bilges. If possible, keep to the same type of paint as used previously, whether red oxide, bitumastic, epoxy, zinc rich or alkyd. There is no need to lug moderate amounts of ballast right out of the ship to paint it – with adequate packing, most pieces can rest temporarily on fixed sole boards and bunks. Wire brushing followed by two coats of paint generally keep iron ballast clear of rust for a few years.

Limber Holes

Once the bilges are dry, start up forward and work back, disturbing all dirt with a paint scraper, old screwdriver and wire brush. A vacuum-cleaner is quite the best tool for picking up the scraps. Without one, a hand brush and small shovel will take care of the bulk, while a tack-rag (or a piece of mutton cloth soaked in varnish) is ideal for final dusting.

Short pieces of stiff wire (some with plain ends and some with hooked ends) are invaluable when clearing dirt from the *limber holes* (fig. 12) which allow bilge water to flow be-

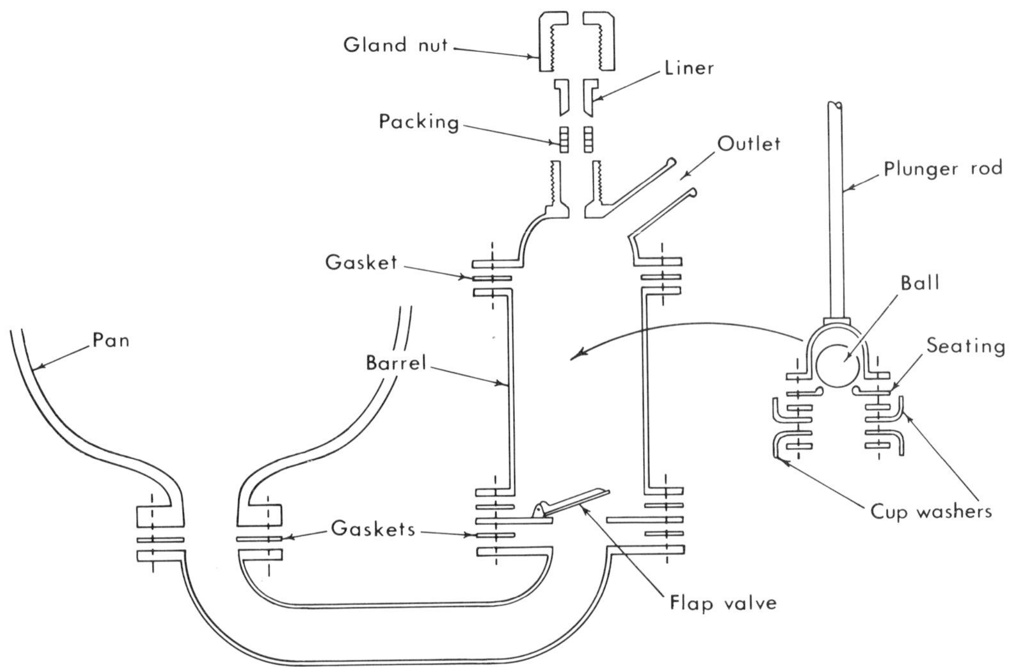

13. Exploded diagram of a toilet soil pump.

neath the frames, floor knees or bulkheads in most craft of wood, metal, fibreglass or ferro-cement construction.

Having removed any small chain or knotted cord rove through the limber holes, scrub these with bilge cleaner and replace only when bilge paint has hardened completely.

Defects

When treating bilges, keep an eye open for defects such as cracked floors, loose fastenings, open seams, corrosion, rot, or faulty skin fittings. Add such items to the repairs list – perhaps for attention the following winter. The same scrutiny is advisable as all interior cleaning and painting proceeds.

You may encounter various electrical connections in the bilges. These could include radio ground (earth); lightning conductor strap from rigging chain plate (or metal mast) to keel bolt; cables from cathodic protection plates (see Chapter 8) to sterntube or gearbox; connections to depth sounder transducer and log/speed impeller; a static suppression lead from a brush rubbing on the propeller shaft. Keep paint away from terminals and label all disturbed wires to ensure that they go back correctly.

Chain Cable

The anchor chain locker is just a space in the fo'c'sle bilges on most yachts. Before cleaning this compartment, all chain must be unshipped and ranged on deck or away from the boat altogether.

If the chain is starting to rust, give affected parts two coats of good aluminium paint. If very rusty, send it away to be regalvanized and tested, or sell it for mooring chain and buy new. If bad rusting has occurred at one end only, replace the affected length with new, using a tested split link the exact size of the existing links. To minimize any slight weakness, make this part the bitter end in the cable locker.

If you intend to keep the ship in first-class condition, this may be a convenient time to send away for regalvanizing any other steel parts showing bad rust, such as anchor, chain plates, fairleads, bollards, mast bands, sheet horse, sheaves, rigging screws or eyebolts.

Toilet

Especially with ladies aboard, a breakdown of the heads (whether motorized or hand pumped to sea, or with holding tank and macerator/chlorinator) can make a cruise tedious – more so when the skipper is not a handyman.

Being basically reliable, such units tend to be neglected, but it pays to take them ashore after about five years' use, for a preventive overhaul. Renew all parts liable to failure (fig. 13) such as pump cup washers and diaphragms, ball and flap valves, gaskets, seals and gland packings. Renew pan and seat if damaged or stained, then clean and paint the remainder. If possible, get a spare-parts list before starting this: sketch any complex parts to make sure they go back correctly.

Sea water does not freeze so readily as fresh, but complete draining is advisable in winter.

Seacocks

Toilet seacocks should be the ones most frequently turned, yet they are often found to be seized in the ON position or so stiff that no one bothers to use them. The big soil pipe

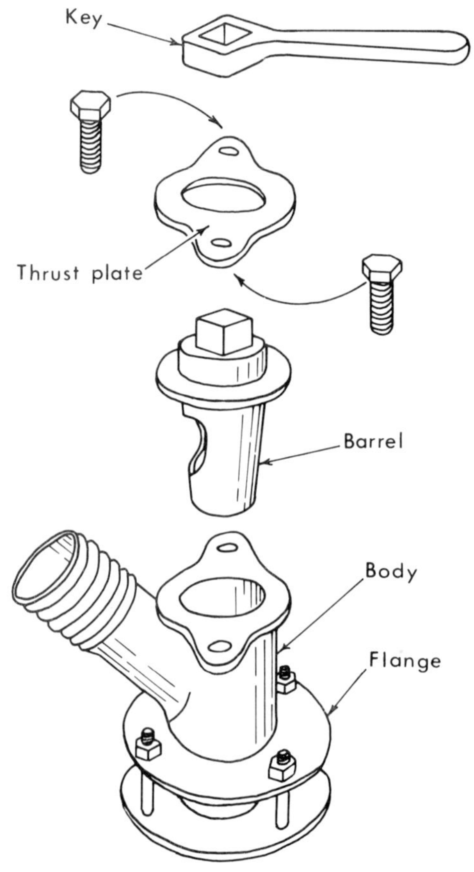

14. Parts of a soil pipe seacock.

valve (fig. 14) is the worst culprit, though easy to dismantle. Polish all working surfaces with Brasso, lubricate with a light smear of underwater (water pump) grease, then re-assemble.

Other seacocks for basin outlets, engine cooling water, deck-wash pump, or cockpit drains, may be small editions of the same type. Some may be gate valves (fig. 15) needing many turns to open and close. The cores of these are readily withdrawn by unscrewing the big hexagonal cap on top of the body, having first opened the valve about half-way.

If there are several valves to deal with, label the loose parts and take them home with you, or to a near-by bench with vice. Take cff the handwheel and clean the stem. Rotate the stem as though closing the valve and, on reaching the end of the thread, slide the gland assembly right off. Clean the parts thoroughly in thinners, then polish stem and screw thread on a buffing wheel, or with metal polish and rag.

If the stem was clean and oily before withdrawal and the valve was known to have a tight gland, it should not be necessary to rake out and renew the gland packing. If the packing was disturbed or appears hard and dry, remove the nut and liner which squash it down, rake it out and fit new packing after having replaced the valve stem. Suitable packing should be available from any plumber.

Gate valves (and mushroom valves of similar appearance) are not only used on skin fittings. You will find them on tank outlets

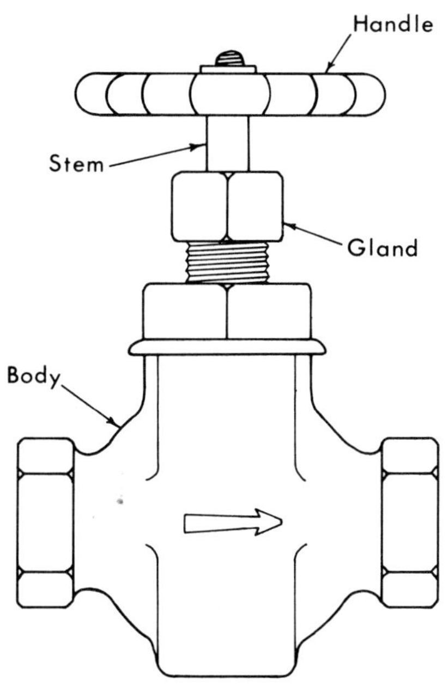

15. A gate valve.

and balancing pipes, also on *pump manifolds*. The latter are sets of valves – perhaps mounted on a single body casting – enabling one pump to handle alternative jobs, such as bilge pumping, bath emptying and deck washing.

Pumps

Most hand-bilge pumps are of the diaphragm, plunger or semi-rotary type, all shown in plate 11. The first two are easy to strip down and then install a new kit of wearing parts – usually available direct from the makers or through chandlery stores.

Semi-rotaries are often the most troublesome and they were intended to be scrapped when worn. However, the amateur mechanic can do a winter rebuild job on one of these at modest expense. Most diaphragm pumps are mainly of plastics with renewable working parts. An overhaul can sometimes be accomplished in less than one hour.

When unscrewing the base of a plunger pump to gain access to the lower valve, be careful not to dent the barrel, whether of metal or plastics. Many yacht toilets are equipped with pumps of the plunger type.

Power-driven centrifugal pumps (as used for water supply to baths and basins) have no valves to cause trouble. With proper attention to gland packing and greaser, they last almost indefinitely without needing a new impeller or bearing. Replacement impellers and gland packings for most vane-type power pumps are usually available from the makers and are simple to install.

Piping

Hose clips on pumps and general piping

Plate 11 (*a*). *Diaphragm pump with one access door open.*
(*b*). *Plunger pump.*
(*c*). *Semi-rotary pump.*

should be replaced if rusty. Stainless-steel ones are well worth the extra cost.

Copper pipe connections to engines tend to become hard and brittle after about three years, due to vibration. To prevent fractures, they should then be removed and softened by annealing – heat to medium redness and allow to cool. A blow torch passed slowly along the pipe will do this, though better results are achieved by heating in a furnace and quenching immediately in cold water. There is generally no need to heat the endmost parts, thus avoiding the risk of loosening any brazed unions.

Engines

With the lavish accommodation of most modern yachts, the engines seem doomed to be crammed into a small space, requiring the services of a contortionist or a midget to carry out maintenance! Winter is the time to attend to serious engine work, as extensive dismantling of obstructing panelling, flooring, bearers and bulkheads is then no great hardship.

Now is the most convenient time for a top overhaul, but if the sump needs cleaning or big-end bearings renewing (and you have a modern unit based on a vehicle engine) it will doubtless mean jacking the engine up or lifting it right out of the boat. The older type marine engine had removable panels in the crackcase side to permit most work to be done without upsetting the alignment.

Lifting main engines from a motor yacht generally means removing part of the wheelhouse roof. The operation is difficult inside a shed and is best done just before hauling out. Small auxiliary engines are often manhandled over packing laid across the cockpit side decks without requiring a hoist. Always check shaft coupling alignment finally after the boat has settled down in the water.

Correctly winterized engines (see Chapter 3) in good order normally need little attention. If not properly laid-up, a few turns by hand every week prevents bores from rusting and valve stems from sticking.

Smart engines usually run best, so give serious thought to cleaning down with thinners and applying a coat or two of the maker's recommended heat-resisting paint.

Heating Equipment

Oil-burning stoves, water heaters, evaporators, air conditioners and other complicated equipment is best serviced by the makers or their agents. With the small amount of use on board most yachts, servicing should not be needed every year, but electrics must be kept dry and motors run occasionally.

Tanks

If not emptied, fuel tanks should be filled to exclude air and moisture. Check for sludge or water accumulations via the drain cocks at least once during the winter. When a lot of rust comes out, a tank is nearing the time for renewal. Small brass tanks showing jelly-like sludge are best unshipped and swilled out with cellulose thinners – not forgetting the fuel pipes and filters connected to them.

Drinking-water tanks of galvanized steel or fibreglass generally have bolt-on circular cover plates. These need removal every three years for internal cleansing and examination by flashlight. When steel tanks rust, renewal is a better bet than regalvanizing. A squashed or hardened cover plate rubber gasket should be renewed.

Formalin is a good harmless mild disinfectant to use when washing out drinking water tanks, but any substance containing phenol could taint the water for a long time. Bilge tanks often rust outside long before the inside deteriorates. Removing them may be difficult, but there is no alternative if they are to be preserved.

Galvanized steel tanks are rarely painted, but when rusting starts (usually at the bottom) coating with epoxy-pitch paint could add a further ten years of life, provided the interior is good. Diesel fuel tanks are rarely galvanized outside and need to be extra well painted. Where such tanks are boxed in permanently, use all possible tactics to get at the hidden parts.

Tanks of resinglass could be damaged by chafe if not correctly mounted; otherwise no external maintenance is required. When such tanks give a foul taste to drinking water, try sluicing a small quantity of vinegar around the inside. Flexible tanks of PVC must lie on a smooth bed. Patching weak places at an early stage is always wise – spare material and the special adhesive required is generally available from the makers of these bags, or through a chandlery store. All water tanks and pipes need draining when frost is likely.

Trailers

A good road trailer or launching trolley should last twenty years if painted regularly and winter is the time to do this.

Wheel bearings and tyres are readily replaced, but once badly rusted, the chassis or framing is a sad sight, doomed to collapse under load one day. Remember that hollow steel box sections and tubes often rust internally. It pays to drill these for screw plugs or rubber bungs, enabling a quantity of used engine oil to be poured inside each winter and sluiced around to coat all surfaces.

Stored outside and often near salt spray, trailers need good paintwork plus occasional touching up to ensure long life and safety.

Dinghies

Unless some repair jobs have been listed, a properly stored dinghy should need no attention until the spring fit-out. Typical dinghy repairs are described in Chapter 6.

Very old varnished dinghies can be rejuvenated remarkably by changing entirely or partly to a painted finish. However, this can prove a lengthy job which must not be left entirely till spring.

Many a fibreglass dinghy has leaky buoyancy tanks. Test these by boring a small hole, pressurizing with an air line, then listening for leaks. Do not fill with water. If leaks are numerous, bore a 2 in (50 mm) diameter hole at one end of the faulty tank, stand the boat on end, and fill with polyurethane foam. This comes in the form of two liquids which are mixed together and poured in; be careful to mix correct quantities for the volume concerned.

Special flexible paint is marketed for treating canvas covered canoes and folding dinghies. Repairing rips in such canvas is difficult. Whether a sewn or glued patch is fitted, old paint is best removed by the careful use of a chemical stripper. Loctite cyanoacrylate glue is ideal for bonding on a canvas patch, but make every effort to apply pressure by holding a pad of wood inside while strutting or wedging from outside. If possible, fit a back-up patch inside also. Finally paint all bare canvas with several coats.

Awnings and covers may be repaired in similar fashion if sewing is not possible. Cotton canvas should be treated with preservative and water repellent – usually obtainable from good marine stores.

five

Spring Care

Dinghy sailors should be able to fit out in a few weeks and be racing at Easter in Britain and the United States. In Canada, thick ice may still be about at that time. In New Zealand – particularly South Island – conditions are similar to those in Britain, but the seasons are almost exactly in reverse by six months.

Most cruising boats get in commission at least a month later than dinghies; only the hardiest souls find weather conditions suitable in Britain before the start of May. This leaves March and April as ideal periods for beautifying the boat, preparing the machinery and re-stowing gear.

Those who did all possible end-of-season and winter chores and repairs will find no difficulty in meeting the deadline. Those who prefer to await good weather for everything may find that, although the hours of daylight increase in spring, other commitments loom up and boating time is all too brief.

Beauty Treatment

Painting and varnishing often takes up more spring care time than all other jobs added together. Over wood, steel and ferro-cement, paint performs a valuable protection role as well as a decorative effect. Over plastics and marine alloy, paint normally serves only as a beauty treatment. However, when such craft are kept afloat, their bottoms are sure to need an antifouling coat to ward off weed and barnacle growth, while sound undercoating helps to prevent the onset of blistering or osmosis damage – see Chapter 9. Many fibreglass boats need topside painting after some six years of hard use if the original beauty is to be retained.

The quality of amateur work should match that of the professional, provided the following pages are noted and makers' instructions adhered to. For good results, surface preparation should take far longer than putting on the paint or varnish. Each process is likely to take an amateur double the time needed by an experienced operator to get similar results. Do not feel bashful about using masking tape (plate 12) to avoid a ragged waterline, plus a sheet-metal shield to cut in corners precisely.

Although light rubbing down, touching up, and one or two coats, constitute the traditional spring beautification, many owners prefer to get the job done just before winter when weather conditions are often better. Certainly that would be advisable if the work is to be protracted by the need to strip off all old material down to the bare hull.

Frostbite (winter) racing can cause surface deterioration to some boats. A certain amount of painting and varnishing is advisable in both autumn and spring under these conditions.

With a boat laid up under a roof or sheeted over, autumn enamel should not lose its gloss appreciably during the winter, and the top coat will be at optimum hardness to resist the chafe of fenders and the shoes of yard hands clambering back and forth across the deck to

Plate 12. Masking tape gives the boot-top line a professional touch.

reach other craft in the spring. The cheap but excellent alkyd yacht paints (see later) take many weeks to reach an acceptable hardness. If applied thickly in cold weather, this period could be doubled or trebled in length. Avoid moist weather conditions at all costs. Finish early in the day to prevent evening dew causing the surface to bloom, necessitating a further coat.

Rubbing Down

Speedy power tools are tempting but, except for the harmless orbital sander, they are best avoided. Belt and disc sanders can create score marks which need the extensive use of trowelling cement to obliterate them completely from the revealing reflections of a high gloss finish.

Hand rubbing with waterproof abrasive paper around a cork block gives excellent results. Dunk it frequently in warm water with a little added detergent. After rubbing, wash down using clean water and a piece of old towel, then dry off with a fresh piece of towel.

A soda block or pumice block used wet behaves similarly. However, paper is more versatile, for when folded and held by the fingers, it will get right up close to beadings, around spars, and into coved grooves.

If there will not be time for complete drying off before coating, rub down with dry sandpaper. This takes longer than a wet rub to form a perfect matt or frosted surface. It also uses about seven times the quantity of paper and smothers everything with dust – including one's lungs when no respirator is worn!

Most amateurs fail to rub down sufficiently. Inspection through a powerful magnifying glass is revealing. The purpose of the cork block is to hit the highspots and miss the hollows, but after a thorough rub, all parts will have been touched to some extent and much of the coating applied last time will have been cut back.

Wet or dry paper (commonly called *wet and dry*) of Grade 180 is about right for the heavy annual rubbing down of mediocre paintwork. On high-class paint or varnish, use the finer Grade 240. To rub a tender surface before applying a second or third coat, use Grade 300 to 400 with light pressure.

Stripping

To repair a hull of metal or ferrocement, and to re-caulk the seams of a planked boat, some or all external paint may need stripping; otherwise delay the evil day as long as possible. Partial stripping (where paint is flaking or cracking) proves quicker than treating the entire hull, though you will need to build back to the original thickness with trowelling cement.

Burning off with a blow-torch and scraper (plate 13) is still the quickest way for over-thick paint build-ups, but this excellent old method has limitations. It might char the wood on varnishwork, delaminate plywood, blister the interior paint on a metal hull, and cook plastics to the point of breakdown. The fumes from burning paint should not be inhaled. Some antifouling paints are particularly poisonous in this respect. With power available, electrically heated torches can be used for burning off.

Chemical strippers soften old paint and varnish layer by layer. It may need ten applications, scraping between each, to get down to

Plate 13. Burning off with butane torch and palette knife.

the bare hull.

On completion, neutralize the action by washing down with either water or turpentine, according to the instructions. Beware when opening a can – pressure may blow the chemical into your face. Goggles make a good precaution at all stages. Keep some of the neutralizer handy and treat splashes on the skin immediately.

Most chemical strippers will damage plastics, but special formulations are marketed to deal with this situation. Otherwise they work effectively and quickly on all surfaces. They are especially valuable when dealing with brightwork.

Mechanical stripping (mainly using a disc or belt sander) is popular for getting paint off plastics. Otherwise this generally proves a noisy and dusty operation.

Priming

After stripping, prepare the surface by rubbing down carefully with dry sandpaper. All paint-makers have a primer to go over each hull material to provide a reliable bond, and be compatible with their finishing paints. They include primers to soak into timber, self-etching ones for marine alloys and plastics, and special formulations for steel, some of which contain powdered zinc. In a few cases, thinned top coat material is recommended as a primer, particularly where varnishes are concerned.

On timber, especially pines, all knots on bare wood should be sealed with shellac varnish, also sold as patent knotting.

After priming, fill any depressions with hard stopping, rub down, then re-prime any bare wood.

A stripped steel hull which was not originally zinc sprayed must be grit blasted and immediately painted for best results. Rusty patches on an otherwise sound surface may be brushed with phosphoric acid (Jenolite) or Naval Jelly four times, wiping off the last coat with dry rag and allowing to dry completely before priming.

On plastics, having rubbed down wet, it pays to rub again lightly with dry paper to provide an ideal key, especially if a proper self-etching primer is not available.

New ferrocement is usually etched with muriatic acid (diluted hydrochloric acid or spirits of salts) before priming.

Paint adheres well to zinc-sprayed steel, but shiny marine alloy really needs grit blasting. If this is not possible, degrease with acetone, give a fine dry rub, then apply self-etching primer.

Choosing Paint

Most boats nowadays are coated with either cheap alkyd paints, or the more expensive polyurethanes. Unless you intend to strip, keep to the same material as before.

To tell the difference, dull a small area by rubbing with fine steel wool, then rub it vigorously with metal polish and buff it with a dry cloth. Alkyd paint will stay dull, while polyurethane will form a high gloss.

New boats are sometimes sprayed with car paint. To test this, rub lightly for about twelve seconds with cellulose thinners. This will dissolve car paint only.

The most expensive epoxy paints could have been used on a steel or ferrocement hull. Most polish up like polyurethanes, and can be over-painted with the latter in any case. Some acrylic paints soften sightly when boiling water is poured over them. Bitumastic paints can be dissolved if rubbed with naphtha.

The alkyds (in common with ordinary boat varnish) last adequately for one season, take a high gloss, and are the easiest to apply. Many owners still prefer them to any other type.

Polyurethanes harden quicker and can be polished to keep a good sheen for perhaps three seasons, making them ideal for those parts of a craft which are unlikely to get chafed or damaged. The one-can variety is air-hardening and not so durable as the dearer catalyst-hardening two-can variety. Peeling is a problem with polyurethanes unless applied under ideal conditions of surface preparation, temperature and humidity.

Epoxies are always two-can and although difficult to apply, they have superb lasting and adhesion properties. They suit steel and ferro-cement hulls best and as they are often made with an egg-shell finish, this helps to conceal the waviness often found in the topsides of such boats.

Special Coatings

Some firms list special coatings for a multitude of boat applications, such as the inside of water tanks, hot engines and funnels, bilges, galvanized fittings, boot-tops, oiled teak, cabin joinery, anti-condensation duties, underwater surfaces, inflatable rubber boats, canvas covers, canvas canoes or folding dinghies.

Coal tar derivatives, such as bitumastic paint and black varnish, make the cheapest of all coatings and are popular for use on old timber and steel boats. They craze with age and are difficult to burn off as they ignite readily. Special sealers are available to enable them to be covered with alkyd paints.

Chlorinated rubber makes a successful and cheap coating for steel while there are special vinyl formulations for steel, for rejuvenating vinyl deck coverings, and for nylon bottom sheathing. The silk-finish vinyl household paints containing fungicides are useful for cabin decoration work as they inhibit the growth of mildew in humid conditions. Make full use of the literature issued by paint makers and do not apply fewer coats than they recommend.

Thinners

In general, each of the products so far mentioned requires a different type of thinners. Diluting paint is not always recommended by the makers, especially for a final coat, but you had better order sufficient thinners in any case for cleaning hands and brushes.

If you happen to run short, it may be useful to know that cellulose thinners (available from car body repairers) will clean out almost any type of paint, so will acetone.

Turpentine or turpentine substitute (white

spirit) stocked by most hardware stores works fine for thinning all alkyd paint and ordinary boat varnish.

Undercoats

Although not essential, it pays to use the special undercoating available for most alkyds and antifoulings. This dries rapidly to give a soft matt finish ideal for a quick rub down. It obliterates well and levels off without brushmarks, ensuring a perfect foundation for the top coats.

It should be applied only after completing all priming and stopping, though small defects may still be made good with trowelling cement on top of an undercoat.

Brushwork

On vertical faces, to avoid *curtains* (run marks) brushing out to an even not-too-thick coating is essential. With alkyds and ordinary varnish this is simple as there is plenty of time to work in big panels, brushing to and fro, then up and down, then diagonally, before laying off horizontally (with slow strokes of a lightly held brush) towards the last completed panel. Standing the can in a bowl of hot water periodically makes brushing alkyds easier in cold weather. Some high-build epoxies have a thixotropic formulation to eliminate curtains even with a thick application.

Using two-can materials, great speed over smaller panels is necessary to ensure a neat join-up before initial setting starts.

Curtailing Dust

Grit and dust are menaces. They can be eliminated from the surface by wiping down carefully with the special muslin tack-rags sold by paint stores. They can be eliminated from the brush by careful brush cleaning, and from paint or varnish by straining through a piece of old nylon stocking. They can be minimized in the air by avoiding draughts and breezes. Damping the floor sometimes helps when using alkyds, but it could humidify the air too much in a confined space when using polyurethanes.

Note that a tack-rag works infinitely better than a piece of cloth soaked in turpentine. Occasional insects do not matter – they may be wiped from the hardened surface later, leaving only microscopic marks.

The gritty all-round rope fendering is easy to remove from a dinghy before rubbing down. There is no need to remove the D-section rubber type (fig. 16) every year.

Varnishing Teak

Before varnishing bare teak, degrease by doping the surface with trichloroethylene and wiping with clean rags. Some paint-makers

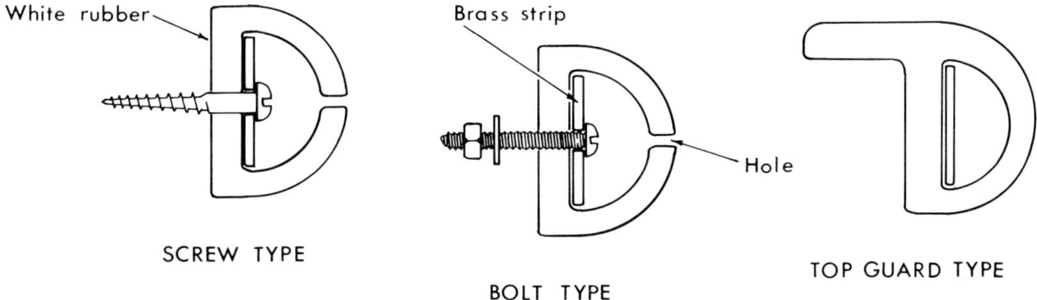

16. Rubber fendering for dinghies.

will supply degreaser, but if you use carbon tetrachloride (sold for cleaning grease spots off clothing) beware of the poisonous fumes. A few firms make special primer varnish to ensure good adhesion to this difficult oily wood, but the standard procedure in all varnish work is to dilute the first coat with 10 per cent of thinners. You will generally need to apply at least three full coats on exposed surfaces. Bare teak and other hardwoods for cabin joinery may be treated with teak oil in accordance with the makers' instructions. Much-handled parts keep cleaner if coated with egg-shell varnish.

Rollers

Almost any material can be applied rapidly and evenly by using a paint roller. The stippled effect matters little on antifouling for cruisers, but for a gloss finish you must have a helper to follow behind laying off with a brush. A brush is always necessary in any case for working into corners and striking in edges.

Moquette and lamb's-wool rollers work best, but foam ones are so cheap that one can afford to slip them off the spindle and discard them after use. Be warned, however, that some styrene-based materials dissolve foam rollers. When in doubt, test it out!

Cleaning Brushes

After a tiring paint job, there is a temptation to dump the brushes in a jar of water till next day. Immediate cleaning always pays, however, and is essential with two-can materials which go rock hard when left, even when submerged in thinners. In fact, one often needs to clean such brushes several times during a long session.

For economy, keep thinners in three cans – clean, twice used and about five times used. When brush cleaning, wipe first with newspaper, then more vigorously with rag. Press the bristles firmly into the dirtiest thinners some twenty times, wipe with rag, then repeat the process in the less soiled fluid and again in the clean thinners. Peel the bristles back to check if the roots are clean. If so, wash twice in warm water and detergent, then rinse and leave to dry.

If work is to continue within a week, omit the clean thinners stage and the washing. Dangle the brush in clean fluid with the bristles completely submerged, yet clear of the container bottom to allow for sediment to form and to prevent damage to the bristle tips. To hold the brush, use a small G-cramp horizontally or bore a hole through the handle and poke a piece of stiff wire through, just resting on the can's rim. Place rag on top to minimize evaporation.

There is one important deviation from this procedure. For some strange reason, dunking a brush used for ordinary boat varnish straight into turpentine causes the varnish to congeal into a million tiny specks. To eliminate this, use a 50/50 mixture of raw linseed oil and turpentine for the first cleaning stage, thereafter using straight turpentine or its substitute.

Cleaning rollers is similar to brushes, using the sloping paint tray to hold the thinners. You will need a great deal of newspaper and rag, plus a lot of time.

To store brushes and rollers, wrap them lightly in paper or rag and keep them in a cardboard box, duly labelled.

Buy only best quality brushes and try to keep a 2 in (50 mm) brush especially for varnishwork.

Antifouling

Some of the dearest (and most effective) antifouling paints, as well as some of the cheapest, must be applied not more than 24 hours before launching. In between come the hard racing finishes which can be applied well beforehand and do not suffer if the hull is left out of the water later on.

It pays the beginner to discuss his needs with a local stockist or boatyard. When changing boats to one of a different hull material, or changing from a deep water mooring to a half-tide one, a change of antifouling type is sure to be necessary.

To avoid galvanic action, beware of using a copper-based paint on the bottom of a metal boat.

Racing enthusiasts may need an antifouling paint which can be burnished several times during a season. The soft matt graphited paints sometimes preferred for racing have no antifouling properties and thus suit boats which are stored ashore between races.

Outdoors

Unless one lives in an ideal climate, fitting-out in the open air is a tricky business demanding some care to ensure that no time is wasted on fine days doing jobs below decks.

Indoors or out, a job list is essential. When working in the open, divide the list into five sections – work for best weather; dry but cold and damp weather; rainy days; jobs to do at home; tools and materials to bring along next visit. Lists are easily mislaid, so keep a duplicate one in a safe place.

In the Water

Spring chores such as bottom, boot-top and deck painting create problems when a boat is lying alongside other vessels. However, other work can proceed unhindered, particularly when loose gear, masts and spars have been taken ashore for overhaul.

Slipping or drying out for bottom and boot-top treatment can well be left a month or so later than normal to await good drying weather. This also means that the antifouling paint will be fresher when weed growth accelerates.

Topsides are quite easily painted afloat by using a pontoon or raft – see Chapter 2. The process takes rather longer than it does when hauled out. When lying alongside another vessel (or tied up to a quay wall) all operations on one side need to be completed and hardened before turning the boat around.

Spars

Anodized metal masts and spars should only need a wash down with soapy water followed by a good rinse to get rid of salt and storage dust. Deep abrasion on gold anodizing should be hidden by touching-up with matching cellulose paint. To get the correct shade, you may need to mix a little aluminium paint with metallic gold lacquer of the same brand.

Alloy spars which were never anodized are best brightened up by using soap-impregnated steel wool saucepan cleaning pads with plenty of water. A coat of the wax polish marketed for aluminium boat hulls will prevent a patina from forming. Transparent metal lacquer will last several years, but is difficult to strip when the time comes – see later.

Traditional spruce spars need good protection. Once the varnish gets damaged or neglected, weathering blemishes the wood. Small areas can be hidden by the careful application of matching paint – usually prepared by mixing teak-brown with yellow.

A darkened and mottled old mast needs stripping, followed by careful shaving with a cabinet scraper. A plane could be too drastic and if this has been done several times before, a racing boat mast could be seriously weakened.

One rarely bothers to remove the mast bands, tangs, winches, cleats and other fittings, so any previous reduction in mast diameter is usually detectable by careful examination at these points.

Painted masts look fine on certain types of craft and it often proves wiser to change from varnish to paint when in any doubt about the advisability of shaving down the wood.

Rigging Equipment

The standing rigging may be permanently spliced around a gaff mast. Suspend the shrouds well away from the stick before starting to rub down.

Most other rigging items can be taken home. Fitting-out is greatly accelerated by working on these at odd moments when tackling the hull is impracticable. The same applies to all the other mobile parts of a boat as mentioned in Chapter 3.

Galvanized steel wire ropes are best smeared lightly with anhydrous lanoline for weather protection. Dunking in thin oil and then leaving to drain is sometimes suggested, but it tends to attract dirt which marks sails and clothing. Although it protects the internal strands well, it does little to protect the outer strands which often rust first.

Salt spray near the deck creates the fastest corrosion, so it pays to turn all shrouds end-for-end (where practicable) at the first signs of rust. Similarly, for maximum life, reverse both wire and cordage running rigging which is constantly nipped around small sheaves or subject to chafe at localized places.

Stainless-steel wire rope, piano wire, or rod rigging, needs no maintenance apart from washing and close examination for broken strands, drawn splices, general chafe (including the insides of thimbles), buckled terminals and loose swages.

Blocks

Modern sheaves and blocks are not normally intended to be lubricated, but a spring-clean with warm soapy water is not a bad idea, helping to remove grit which causes wear and strange noises under load.

Traditional blocks are much cheaper for big craft, but they do require lubrication and other maintenance chores. They are ideally suited to gaff-rigged boats and the characters who sail such picturesque craft rarely object to the annual ritual of overhauling a couple of dozen wooden-shelled blocks!

The basic tasks normally proceed as follows:

1. Drive out the pin, sometimes held captive between inlaid metal discs held in place by four small woodscrews (see fig. 17).

2. Remove the sheave and rub down between the cheeks as well as outside the block.

3. Dab two coats of varnish on any bare wood, followed by one coat all over.

4. Brush sheave and pin in paint thinners

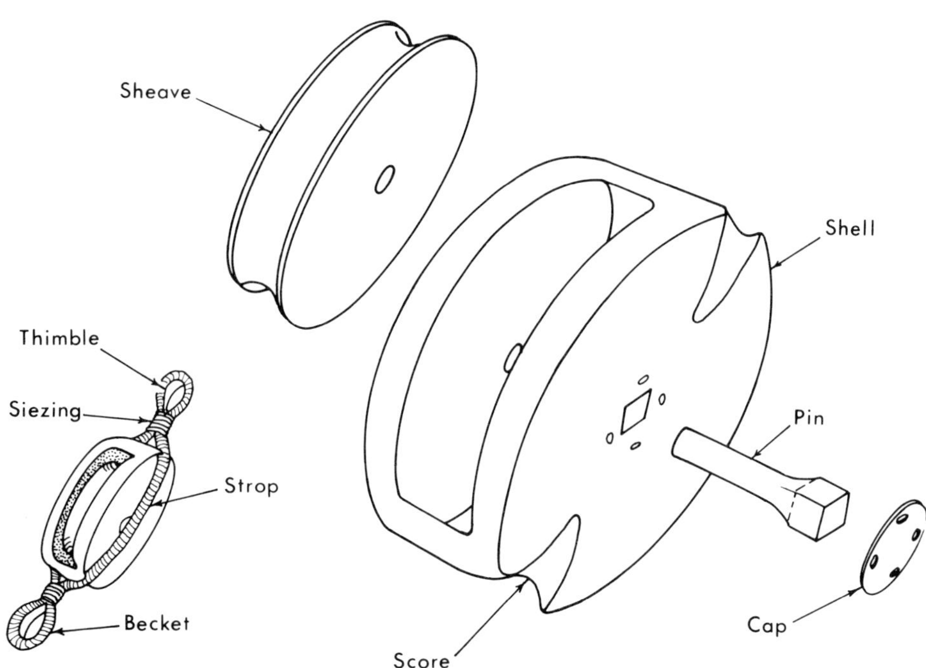

17. Dismantling a stropped block.

and, when dry, lubricate both with underwater grease (water pump grease) and reassemble.

5. Rub down lightly outside and give one further coat of varnish.

When a block is stropped as in fig. 17, the seizing must be cut away to slacken the strop and allow the pin to be removed. This also enables both shell and strop to be properly varnished. After replacement of the strop, give the new seizing (whether of wire or cord) a few coats of varnish.

An internal-bound block has no strop – the eye and becket are part of a galvanized steel cage encased by the wooden shell. The visible metal parts are normally coated with aluminium or black paint.

A worn pin may need renewal, though building up with weld (followed by grinding down to original size) saves making or searching for a replacement. Turning the square end 180° solves the problem of a pin worn on one side only.

A worn modern block of Tufnol and stainless steel should be serviced by the makers long before a slack sheave bottoms in the shell and jams.

Having overhauled all running and standing rigging, its installation is called *dressing* the mast. Generally speaking, this job is best left until a short time before launching. Varnish will then have had a maximum time to harden and a minimum of dust and grit will land on lubricated parts.

Underwater Fittings

When treating surfaces below the waterline, avoid clogging intake grilles with paint, also small openings such as the water feeds to cutless rubber bearings on propeller shafts. On the contrary, large open pipes (such as toilet soil outlet and cockpit drains without rubber flap valves) are best kept free from obstructing barnacles by brushing antifouling paint well inside them.

Sacrificial zinc anodes (plate 14) must be left bare; so should a radio ground (earth) plate and a depth sounder transducer. Temporary masking tape is advisable on these. If a barnacle ever does cling to the transducer, dissolve it by applying hydrochloric acid (spirits of salts) with a small brush – scraping could easily cause damage.

Copper sheathing needs no paint. The same applies to a keel cooler for engine water, if the tubes are of copper. Some paint on the tubes is generally unavoidable when coating the planking behind, but the outer pipe surfaces are easily wiped clean. Brass tubes and rudders are not immune from barnacle growth and are best antifouled. Propellers remain free when worked regularly. Special paint is made for them, used mainly on big outboards and stern drive units.

Rust

Corrosion pitting in thin steel hulls can eventually cause leakage. Occasional holes may be effectively sealed with fibreglass or saucepan menders! See Chapter 8 for further details of such repairs.

On thick steel or aluminium plating, isolated pitting is easily checked by cleaning up with a burr drill and applying epoxy putty. Large unpitted rusty areas should be scoured with a power wire brush, doped four times with phosphoric acid (Jenolite), wiped with rags while still wet, allowed to dry (perhaps urged on with a blow torch or hair dryer), given two coats of zinc paint, then built up

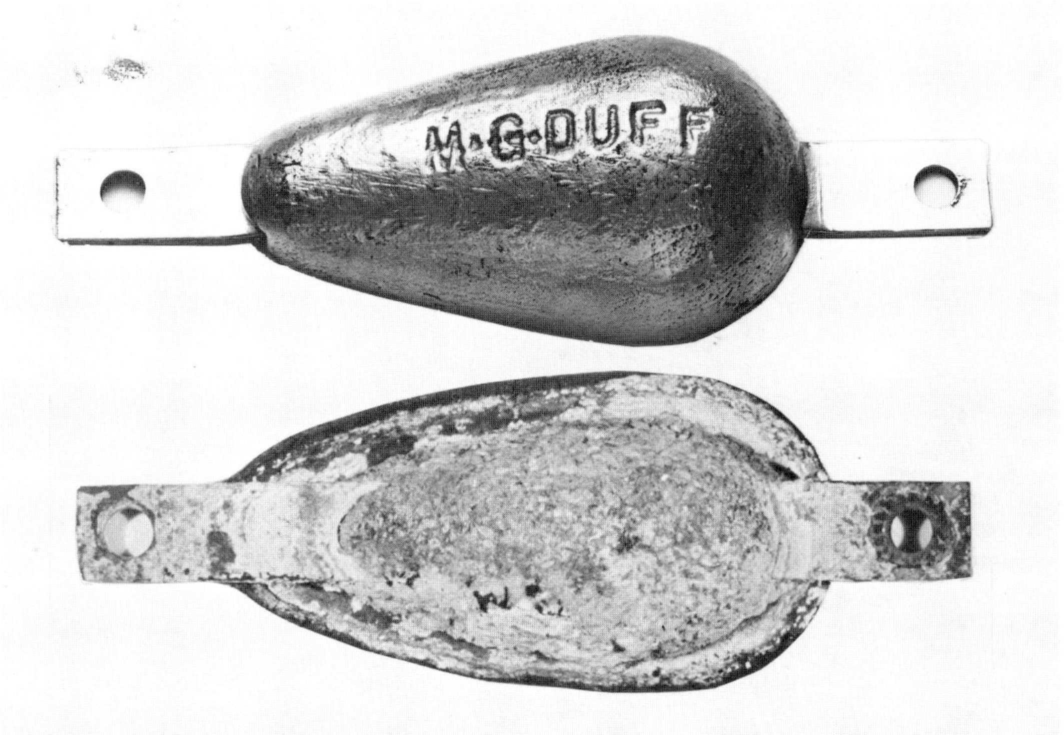

Plate 14. Before and after views of sacrificial anodes.

with the boat's normal painting specification.

Similar treatment copes with rusty steel chain plates, rudder, centreplate and bilge keels. Small steel deck and mast fittings are often finished off with aluminium paint (to resemble their original galvanizing) but other colours sometimes look better.

When rusty fittings have been re-galvanized (see Chapter 4) paint will adhere better if the galvanizing is allowed to weather for a year: the same applies to a new boat with galvanized fittings.

Polishing Brass

Few modern boats have polished fittings of brass, copper, bronze or gunmetal – they have been superseded by plastics, anodized aluminium, stainless steel and chromium plate on whitemetal.

Once neglected, brasswork requires patient restoration, but the operation is greatly simplified if the parts concerned are unshipped for treatment at home.

Most strong acids will remove verdigris. Submersion for a few days in weak acid generally works better than instant pickling using strong acid applied by brush. Hydrochloric acid (spirits of salts) works fine. Sulphuric acid (used in batteries) is easier to obtain, but the concentrated form can be dangerous in inexperienced hands.

Strong alkali (such as caustic soda or oven cleaner) will remove the patina on brass if given time, while the mildest treatment – concentrated ammonia solution with vinegar or table salt added – is useful on parts which cannot be removed from the boat. Strong chemicals could blister surrounding paint or varnishwork.

Once pickled, the next process is polishing. Although it will not reach into all crevices, the best tool to start with is a standard buffing mop used on an electric drill or bench grinder. Such a mop consists of numerous discs of thick linen packed together like a wheel. Sticks of buffing composition are available from tool stores. A touch with one of these on the periphery of the spinning mop every ten seconds will ensure fast polishing.

A fitting with rough or pitted surface may need attacking with dry emery tape before buffing. If coarse emery is used, be sure to remove all scratches by changing to grades of increasing fineness. For a bad surface, start with Grade 1, then use Grade 0, followed by Grade 00.

Rubbing with household metal polish on a soft cloth soon produces a good lustre, but this has to be repeated many times to obtain perfection. Strips of cloth (such as hems torn from old items of clothing) doped with metal polish and pulled to and fro across rounded surfaces (for example, cleats, fairleads and ventilators) produce good results in quick time.

Successful lacquering can eliminate further polishing for up to three years, but the correct materials are rarely stocked by shops or marine stores. Yacht varnish is a poor substitute – its deep colour dulls the brilliance of bare polished metal.

Aerosol lacquers designed for protecting car chrome will do a passable job, but if you have an old craft with attractive brasswork, it pays to keep at least some of the parts polished weekly.

Nameplates

The torment of having to paint a big registration number on the bows of an attractive yacht is best minimized by ensuring neatness.

The transfers sold by marine stores are not cheap and generally need renewing each year. Lay them in varnish with a further coat on top. Rule a faint pencil line on which to align them.

Self-adhesive or screw-on numbers and letters in plastics and metal never look as neat as painted lettering or transfers, but most can be removed and replaced several times without much trouble.

Varnished name boards look good on a brightwork backing, particularly with gold-leaf or gold transfer lettering. Nameplates on painted transom or topsides look neatest when coated with matching paint before lettering.

The cheapest lettering is done by borrowing a set of zinc stencils from a builder or sign-writer, finally filling in the blank bits with a small brush. If an aerosol is used, get synthetic paint – cellulose car paint is liable to blister ordinary topside enamel.

To help foil thieves, carve the name into the transom of a wooden dinghy. A quicker way (best suited to an old boat) is to borrow a set of branding letters and burn the name on!

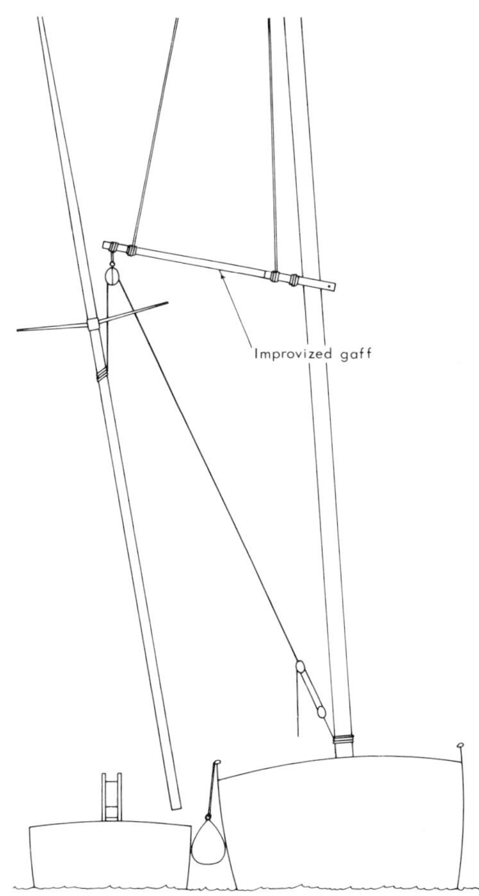

18. Improvized gaff for mast stepping.

Launching

Although many small tasks are bound to remain after launching a boat which has been laid up ashore, getting afloat usually marks the end of spring care.

For a big vessel, launching is normally carried out by a boatyard and is just the reverse of the hauling out procedure described in Chapter 3.

The amateur working alone and launching from a mudberth or high beach may need to link the occasion with an exceptionally high tide. He may also need to dig a trench for a short distance from the keel to make sure she moves off. Leaving out all internal ballast is often a big help – also facilitating the detection of unexpected leaks.

Stepping Masts

When no yard crane or quay is handy for stepping masts, a certain amount of engineering or Boy Scout know-how is useful.

To obviate the need for a helicopter, one of the simplest and safest alternatives is to get help from the owner of a really big sailing craft. Float the mast out to this boat – attaching buoyancy where necessary – then hoist the mast from the end of an improvised gaff, raised by halyards the correct distance up his mast (see fig. 18). Hand-held guy ropes are adequate to keep the gaff outboard, but it pays to complete the operation early one morning before the powerboat cowboys start disturbing the water surface!

six

Dinghy Damage

Often herded together at landing stages, left swinging wildly at moorings, towed by parent vessels or hoisted on board, general purpose dinghies – whether of the rigid or the inflatable type – are lucky to get through a season unscathed.

Racing boats collide at speed, are shifted by gales in dinghy pens and may be scraped by sand or pebbles when launched and recovered in the surf.

Inflatables

Mainly due to their light weight, ease of stowage and good resistance to collision, inflatable boats are widely used as tenders, sometimes fitted with powerful outboard motors.

However, they do have disadvantages. Good quality ones are expensive and they generally need to go back to the makers when damaged. They are pigs to row and often need partial or complete deflation to stow on board neatly. They have little freeboard and almost no space for bilge water. When lightly loaded, a strong wind can flip them over.

Serious damage must be repaired professionally, but pinholes and small tears are easy to mend at home or on board, provided the correct kit is available. Follow the instructions carefully, for a patch needs to be correctly done first time. In attempting to remove a temporary patch the boat's skin could be damaged.

Carry spare bellows, for these are indispensable and difficult for the amateur to repair even temporarily.

Rigid Dinghies

Despite the popularity of inflatables, great numbers of wood, metal and plastic dinghies are in use. The robust and attractive clinker (lapstrake) type is making quite a comeback, proving especially cheap and interesting for amateur boatbuilders to construct. Such craft are the best of all for rowing, but lightweight versions with glued clinker plywood planking and no ribs are popular for car top stowage and carrying down beaches. Such a dinghy weighs no more than the equivalent size in plastics or aluminium. A fully-framed plywood dinghy can be the heaviest of all. Frameless plywood dinghies are often very light, but also tend to be fragile.

Generally, the need for repairs is equally frequent for all types of construction, but the amateur handyman is usually happier working with wood than with the more modern materials.

Invisible Repairs

When a dinghy gets damaged, the culprit (or your insurance company) may agree to pay for professional repairs. If the boat is old or neglected, you may have to stand part of the cost yourself, as her condition may affect the

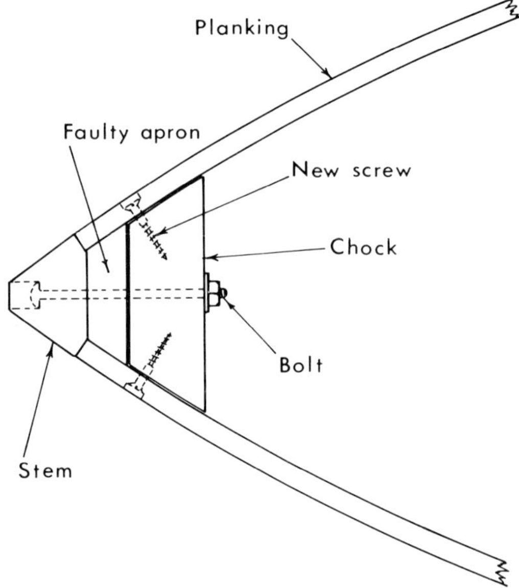

19. Reinforced stem fastenings.

extent of the damage. One usually has to pay a proportion of any insurance claim and a good premium discount may have accrued. Under these conditions, it often pays the able owner to repair collision damage himself as well as taking care of normal wear and tear replacements.

One can either repair to ensure strength and watertightness, or take more time and care to conceal any scars. Almost any part of an open boat can be seen and if she is of varnished wood, extra care is necessary to choose the right timber and to stain this correctly before re-varnishing.

When matching proves difficult, you may have to replace the whole length of a gunwale rubber or coaming instead of letting in a small piece. Fortunately, one cannot see both sides of a boat at the same time! On the contrary, if the *breakwater* (spray deflector) across a sailing dinghy's foredeck gets damaged (see later) it may be wise to renew both halves to ensure an invisible repair which will not lower the boat's value.

Few amateurs are capable of patching the skin plating of a metal dinghy invisibly inside and out, but this is comparatively easy for all types of wood construction, especially when the surfaces are painted rather than varnished. Fibreglass patching invariably shows a bulge on the inside, but thanks to the natural roughness (plus some matching paint) no great harm is normally done to the appearance.

Stem

Given time, even the stem or keel of a wooden boat can be renewed invisibly. Although not practicable for a class racing dinghy, a torn out transom is sometimes best repaired by fitting a new transom a bit further for'ard, then trimming off all broken plank ends.

A severe bump on the stem can cause many *hood end* fastenings (securing the planking or plywood skin flush into the stem) to break.

Should internal rot prevent additional fastenings from holding effectively, it may be necessary to bolt or screw a new hardwood chock to the back of the stem inside the boat (fig. 19) and drive fresh plank fastenings into that. Sometimes the tapered nose of the stem (*cutwater*) gets so badly damaged that the only cure is to remove all through-bolts, saw the old cutwater off close to the planking, glue on a new piece and then replace the bolts.

Resin Putty

The amateur can often save much time and money when repairing a hull built from any type of material by utilizing the range of resin putties now available. The popular ones used for filling dents and small holes in car bodies are equally applicable to boat hulls of metal or plastic, while special ones are marketed for building up wood and ferrocement. To bridge gaps (and in some cases to improve structural weaknesses) laminations of fibreglass – perhaps finished off with resin putty – are again a possibility with all types of hull material.

Whatever the material, the problem is getting any sort of resin to bond firmly. Epoxy resin products are normally more reliable in this respect than the cheaper polyesters used in moulding hulls, but surface preparation must be carried out meticulously with sufficient roughening to remove all impurities and to provide a good key.

When applying glass laminations, woven glass cloth is generally stronger and easier to handle than the chopped strand mat used in moulded hulls. Each successive layer must overlap the previous one by about 1 in (25 mm) around the edges.

Epoxy resin putty activated by a separate catalyst (hardener) is stocked by good marine stores and should always be used in preference to the car and household variety. When in difficulty, a reasonable putty can be made by mixing shredded glass strands with resin to the consistency of thick cream. Although you can buy shredded glass fibre, small quantities are readily made by shearing chunks about $\frac{3}{8}$ in (9 mm) long from the glass fibre rope known as *rovings*. Oddments of the cloth called *woven rovings* will do just as well.

You can make a smooth putty for finishing work by using French chalk or talc as the filler. A paste catalyst mixes better than a liquid type to activate putty and is usually coloured to show evenly throughout when correctly mixed.

Too much hardener causes fast setting – especially in hot weather – so do not waste material by mixing too much at one time. Storing the unactivated putty in a refrigerator until needed is often a good idea. When catalyzed, cold resin remains workable for a long time.

To ensure a good bond to metal, wood, or fibreglass, neat activated resin should be brushed thinly over the surfaces and the putty applied just before the resin gels. On most woods, two coats of resin are advisable, the priming coat soaking in and setting before applying the next layer.

Resin Glues

Although glue helps considerably towards the strength of a joint, gluing to old weathered timber rarely works well. However, glue generally seals a joint against the ingress of moisture better than paint. In general, the dearer the glue, the better the bond.

Resorcinol resin is the most expensive waterproof woodworking glue and is the only reliable one to use for teak and similar oily timbers. With this type, hardener and resin are normally mixed before use and the air temperature needs to be high for best results.

The cheapest resin glue is one-shot (such as Cascamite) which does not need a hardener. The white powder is mixed with water to the required consistency and then applied to both surfaces of a joint. Fresh material gives excellent results, but old stock needs testing by making a small sample joint which is then broken apart. With sound glue, the wood itself will fracture before the glue line fails.

The most popular resin glue consists of a syrup (often made when needed by adding water to a powder) and a separate acid hardener having the appearance of water. When gluing a joint, the syrup is applied to one surface and the hardener to the other: once they are cramped together, setting starts. This glue is preferred to one-shot for important joints.

Most resin glues have good gap-filling qualities. In fact, joints are best not cramped too tightly. A close glue-starved joint is often weaker than a gap-filled joint.

Scarf Joints

Pieces of timber can be joined together end-to-end with little loss of strength by using glued diagonal scarf joints (fig. 20*a*) with or without fastenings.

Although traditional clinker dinghies are built without glue, when it comes to repairing them, the use of glued scarf joints (and curved parts laminated from thin strips) reduces weaknesses and saves wasting time hunting for materials.

Gunwale Damage

Without decking to strengthen it, the gunwale of an open boat is easily damaged by crushing, abrasion, or collision. An all-round fender or rope, D-section rubber, or kapok filled canvas, protects the gunwale and topsides splendidly and prevents the dinghy from scratching other boats. If you come into possession of a rowing boat, launch, or runabout with inadequate fendering, put this job at the top of the repair list before the boat is used as a tender.

As well as having a batten inside or outside the planking running the full length of the

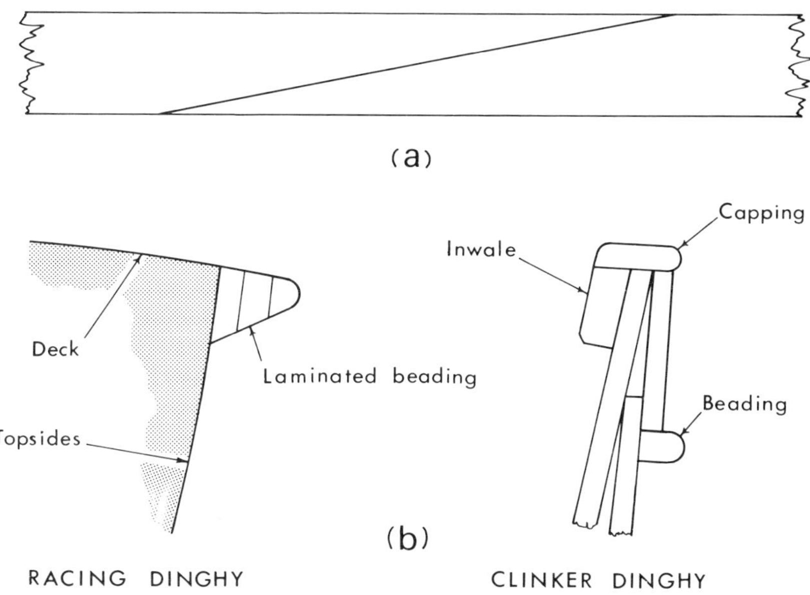

20 (*a*). *Simple scarf joint.*
(*b*). *Dinghy gunwale beadings.*

sheer, a dinghy gunwale may be reinforced with some sort of external beading, see fig. 20*b*. Without the protection of fendering, this is the part most frequently in need of repair.

When this beading is loose, additional screws (or replacing the original ones with the next size thicker) may suffice. When splintered, perhaps by a downward blow, the broken pieces can often be parted neatly with a trimming knife or dovetail saw, then glued back into place. Although G-cramps will fit over the gunwale of an open boat, fine panel pins or small screws may be needed on a decked dinghy. As the original beading will doubtless have been bent into position, some force is often necessary to hold back a broken piece for gluing.

Use cardboard pads to prevent cramps or

temporary pins from bruising the wood. Bore tiny holes for panel pins or they are sure to cause splintering near the ends of the piece.

Boring and Dowelling

Although driving fastenings into softwood is much easier than into hardwood, some sort of pilot hole is certain to be needed with nails and screws in most boat repair jobs.

The depth and size of pilot hole for nails depends on the sort of timber and the type of nail. A few twists of a larger drill may be necessary if the head is to be recessed. Barbed ring nails and traditional copper boat nails need full length pilot holes in hardwood. It pays to make a few test samples, but as a rough guide, make pilot holes for nails at least half the nail diameter, or half the width across corners for square copper boat nails.

Important screws sometimes need four sizes of drill to ensure perfect driving. If the head is to be countersunk you are bound to need three hole sizes, though the smallest one can often be formed with a bradawl. Having judged the drill sizes required for shank and thread-root of a screw (by aligning them while held in the hand) make some tests to determine how much smaller the holes can be made without risk of shearing the screw or damaging its slot. Do not necessarily make the hole to recess a screw the same size as the screw head – the drill may have to match standard dowels or plugs. Even when these are

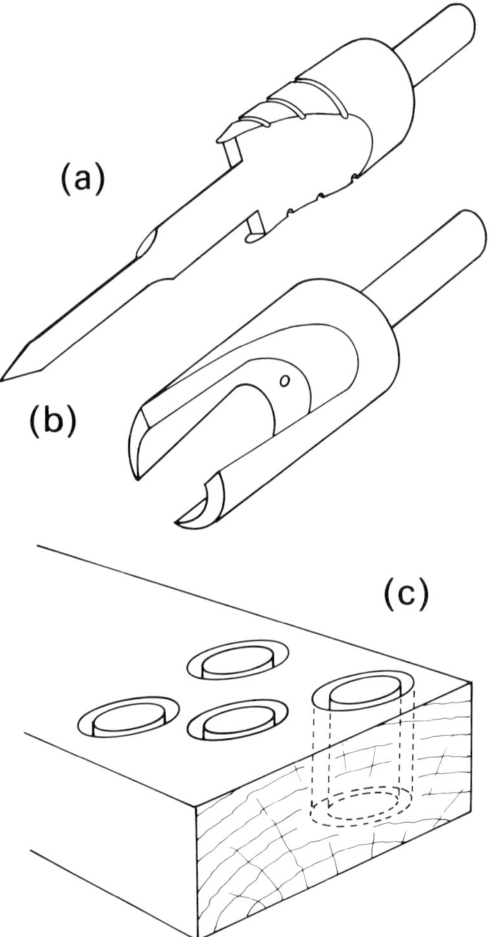

21 (*a*). *Combination drill.*
(*b*). *Plug cutter.*
(*c*). *Dowels cut across the grain.*

not required, make the hole big enough to clear the blade of your best screwdriver.

Combination drills (fig. 21*a*) capable of boring all three holes in one operation are readily available. Most of them overheat badly in hardwoods and they are thus best suited to the final clearing of undersized holes bored with ordinary twist drills. Note that all swarf must be cleared from the flutes of twist drills continually to prevent overheating by friction.

Plug cutters (fig. 21*b*) are generally made an exact match for combination drills, enabling dowels to be cut from the correct timber to camouflage each screwhead. Hard stopping is quicker on a painted finish, but dowels are essential for invisible repairs on varnished parts.

Remember to cut dowels across the grain (see fig. 21*c*) in the opposite direction to the dowelling (pegwood) sold by hardware stores. Stick dowels in with varnish or glue, rotating them until their grain aligns with that of the surrounding wood. Once set, cut off the surplus with a single chisel and mallet blow, then plane and sand flush.

Dowels set in varnish or weak glue are easy to remove whenever necessary by boring a pilot hole in the centre and driving in a steel woodscrew.

When the point comes in contact with the buried screw head, further driving lifts out the dowel. This will not work with shallow dowels, but these are easily broken up with a small chisel.

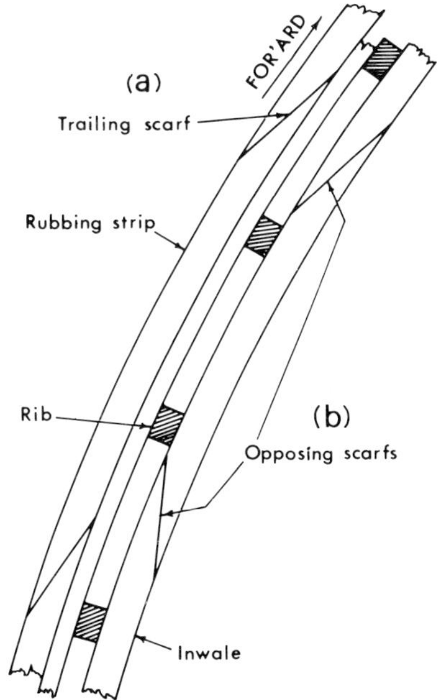

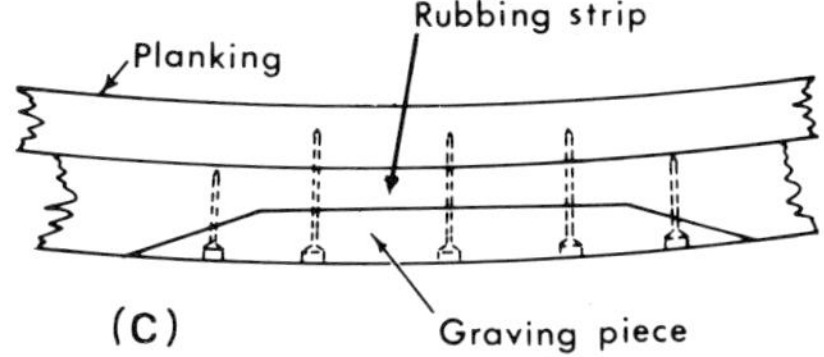

22 (*a*). *Trailing scarfs.*
(*b*). *Opposing scarfs.*
(*c*). *Small rubbing strake repair.*

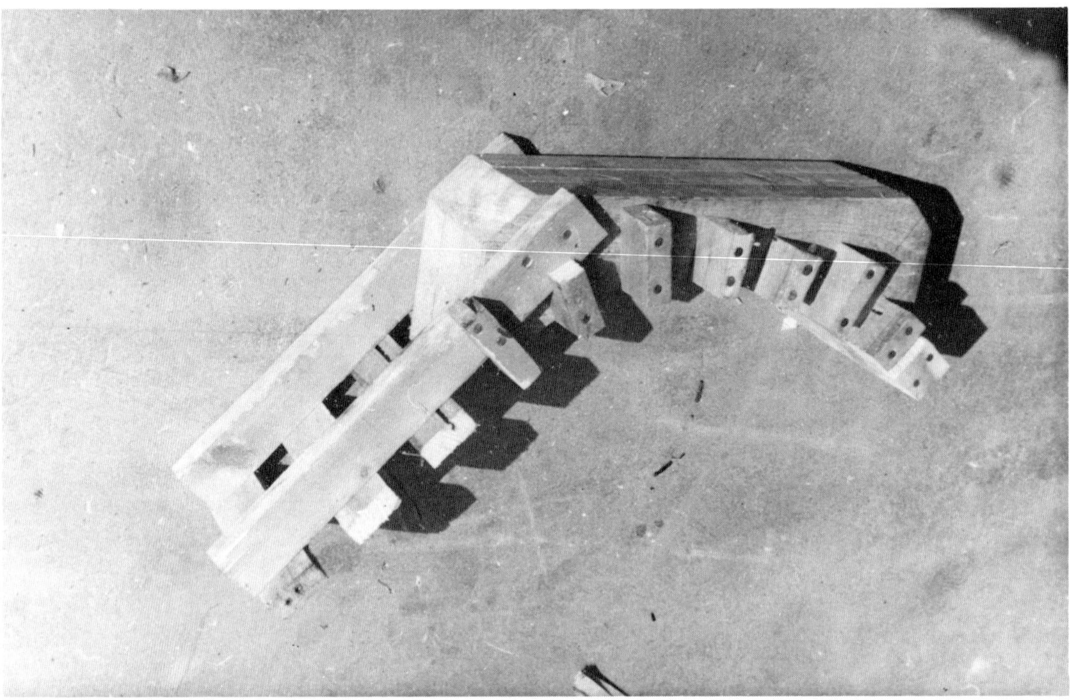

Plate 15. Rough jig for cramping small steamed parts.

Templates

For serious general gunwale damage, scarf in a whole new piece. For localized damage on varnished wood, extend this new piece considerably in either direction. Two scarfs close together rarely help to achieve an invisible repair. Also, remember that a long piece is much easier to bend than a short one.

Scarf joints on external gunwales and rubbing bands are traditionally made to trail aft as in fig. 22*a*, but fitting is simpler for an inside gunwale (*inwale*) if the scarfs face each other as in fig. 22*b*.

To avoid the risk of cutting the new piece too short and wasting it, make a *template* (pattern) of the shape (including the scarf joints) from a piece of card, scrap wood, or hardboard.

Before the repair piece can be marked out accurately from its template, it may need shaping or bending. Cramping one end in the vice and arranging a strut or two from the shed wall will bend most thin pieces.

For a stiff member, make a rough jig as shown in plate 15 and apply G-cramps. To make a permanent bend, certain stiff parts will need steaming or dunking in boiling water for a while before cramping. Apply a little extra bend, as there is usually some spring-back on release. Leave cramped for at least a day to set and dry out.

Dents

Damage from chafe and collision does not always extend right through a rubbing strip. Using the correct timber, a small *graving piece* may be let in (held by glue and brass pins duly stopped over) as shown in fig. 22*c*.

Similar dent repairs are commonly required on stems, rudders, centreboards, keels, thwarts, transom tops and coamings. Work is simpler on painted parts, as fibreglass or resin putty will often suffice. However, even then some scarf-like undercutting at the ends is advisable to stop the repair from falling out one day. A few small nails or screws left protruding a short distance to become embedded in the patch help prevent future looseness.

Graving in the middle of a flat surface (such as planking or one side of a rudder) is a rather different procedure, described in Chapter 9.

Breakwater

Replacing the breakwater, across a dinghy's foredeck creates something of a test for the amateur shipwright's skill! These parts are often of wood even on fibreglass boats, screw fastened from beneath the deck, or glued to strips of quadrant moulding which are fastened separately to the deck.

In the unlikely event of the damaged piece coming off cleanly, it could make a template from which to mark out the new one. Failing that, whittle away at the edge of a strip of hardboard until it fits snugly on the deck, then use this as a template.

To eliminate the guesswork, use a *spiling block* as shown in fig. 23. While the template (or the actual piece of timber to be used) is held firmly in position across the deck, hold a pencil on top of a small block and slide both from one end to the other, making a pencil mark parallel to the deck surface. If the height of the spiling block is made equal to the greatest gap between deck and template (or timber) edge, accuracy is increased and the minimum amount of timber is wasted when cutting out. If the fit is still not perfect, mark any high spots and shave these down patiently.

Remember, there will be a bevel on the

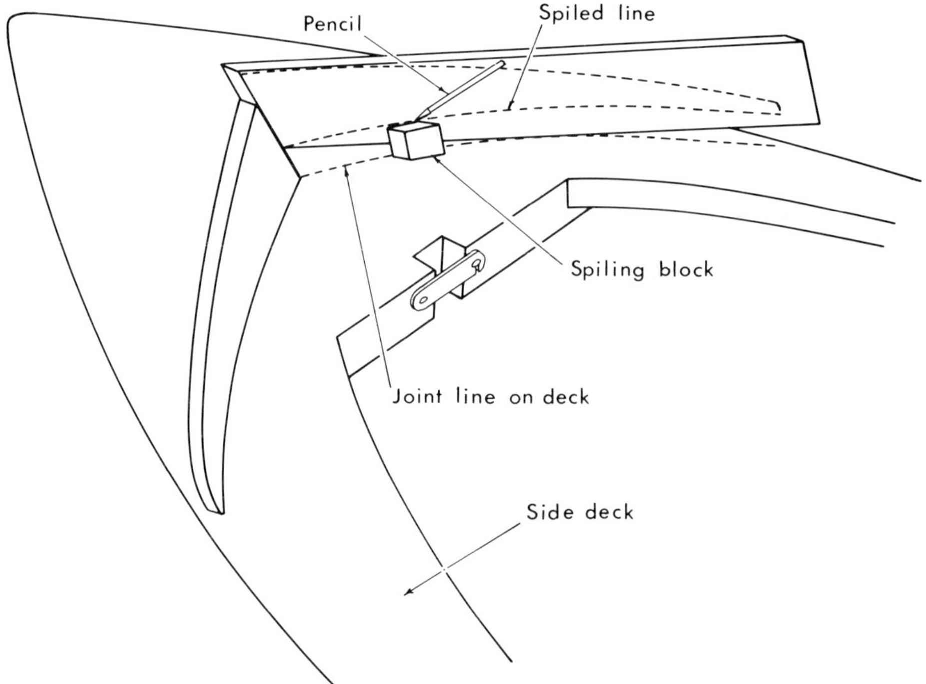

23. Marking the shape of a new breakwater.

bottom edge of the breakwater if it has the customary forward lean. Cut a cardboard template of this bevel and use it to align the saw blade when cutting the curve. As such a bevel is rarely constant throughout, some spokeshave work is certain to be necessary to form a good enough joint for gluing. Do not waste time shaping the remainder of the breakwater until this lower edge is correct.

When the old breakwater was glued to a plywood deck, avoid tearing away the upper plywood lamina. Saw along close to the deck, then plane and sand the remaining stub.

A good glue joint will ensure no leaks via the screw holes. If the foredeck forms the top of a complete buoyancy box (preventing access for screwing) use glue only and fit a reinforcing quadrant moulding along one or both sides of the breakwater. To facilitate cramping, glue the mouldings to the deck first, with a screw at each end. When set, glue the breakwater between them and cramp it by resting bridges of short battens on top, held down by heavy weights.

Coamings

Sailing dinghies usually have thin vertical hardwood coamings attached to the side decks (and sometimes across the edge of a foredeck or after deck) to form backrests and to add stiffness. Renewing a complete section is not difficult, especially if no glue was used originally. With great curvature, laminate the new coaming in situ from two or three thin boards to simplify bending. If you find difficulty in borrowing the many cramps needed, makeshift cramps are soon made from pairs of battens with a bolt through the middle and wedges at one end.

Scarfing in a short section is comparatively simple on straight coamings. Isolated damage caused by vandalism (or some timber fault such as a dead knot) can often be obliterated by inlaying an oblong or diamond-shaped graving piece – see Chapter 9.

When coamings are used as hand-holds for lifting the boat, the wood may split or the fastenings may draw out. Even a severe crack is often reparable invisibly by prising it open, inserting glue with an old hypodermic syringe, cramping tight, then wiping away surplus glue.

Re-fastening with over-sized screws is not always as easy as it appears – dirt in the joint sometimes prevents it from going back tight and the backing piece (carline or beam) could be too thin for secure screwing. In the latter case, small countersunk bolts make a good alternative.

Coamings are frequently secured entirely by bolts on hulls of plastics or metal. Remember that the quadrant moulding usually fitted along the corner between deck and coaming adds strength besides preventing water from seeping into the joint. With the varying angles involved, getting a sound gluing face for the moulding is difficult. It generally pays to bed in mastic and tighten down gradually using long thin screws duly stopped over on completion.

The heads of main fastenings securing varnished coamings to carlines are nearly always hidden beneath dowels. The way to withdraw these without damaging the surrounding timber was described earlier in this chapter, but do not attempt this until all varnish has been scraped from the surface.

Decking

Few wooden sailing dinghies have decks of any material except plywood, and many fibreglass hulls are decked in similar fashion.

On an old boat, the outermost plywood veneer may bubble up here and there – often due to continuous dampness inside built-in buoyancy tanks beneath the decks, plus heat from the sun outside. Some owners forget to remove the tank inspection ports regularly to permit drying out.

Rectifying this neatly on a varnished deck is almost impossible – try cutting out the faulty veneer with a sharp knife, flush over with resin putty and then change from varnish to paint.

When joints on plywood panels break away, refastening is simple, but there is no guarantee against future leakage. When there is extensive joint failure it pays to remove the panel completely, clean off all mating surfaces and re-glue.

Repairs to large and small hulls in plywood are described in Chapter 9. Although access to deck or hull skin for repairs is difficult in way of buoyancy tanks, there is the added advantage that internal patches will not show. Inspection ports are usually big enough to enable temporary struts to be inserted while patches are being fitted. When in difficulty, installing a bigger port sometimes saves the situation.

Plywood Panels

Renewing a complete deck panel means removing all fittings and adjacent beadings. If the old panel is still firmly glued, avoid damage to the under-deck framing by sawing away as much plywood as possible, then stripping the remaining glued pieces with chisel and rebate plane, extracting old fastenings as you go.

Some canoes and racing catamarans have decks with extreme camber, a new panel for which may need steaming over a jig before fitting. Most dinghy decks have little curvature, enabling one to mark out a new sheet while bent in place on the boat.

Shape the inboard edge to a good fit, using a spiling block (fig. 24) if necessary. Then shape the ends, scribing from underneath with a pencil. A rebate width outside such lines may be added from measurements. Alternatively, a little paint applied to the edges of neighbouring plywood panels (or recesses) will leave precise cutting marks when the new sheet is lowered carefully into position. If you are anxious about it, make a hardboard template first. This is especially helpful when the top of a buoyancy tank forms a well-deck below gunwale level, as in the *Mirror* class.

Dealing with plywood on topsides and bottom is almost identical to the procedure for decks, but the curvature encountered in some places makes it advantageous to remove the old panel intact, to form a pattern. If not possible, make a hardboard template. Without steaming, a full sheet of plywood may not bend readily enough to allow direct marking out. Hardboard bends easily in both directions though the contours it assumes are not necessarily the same as those of thicker plywood. Once cut to shape, a plywood panel bends far more easily and even the difficult ones can often be coaxed into place without need for a

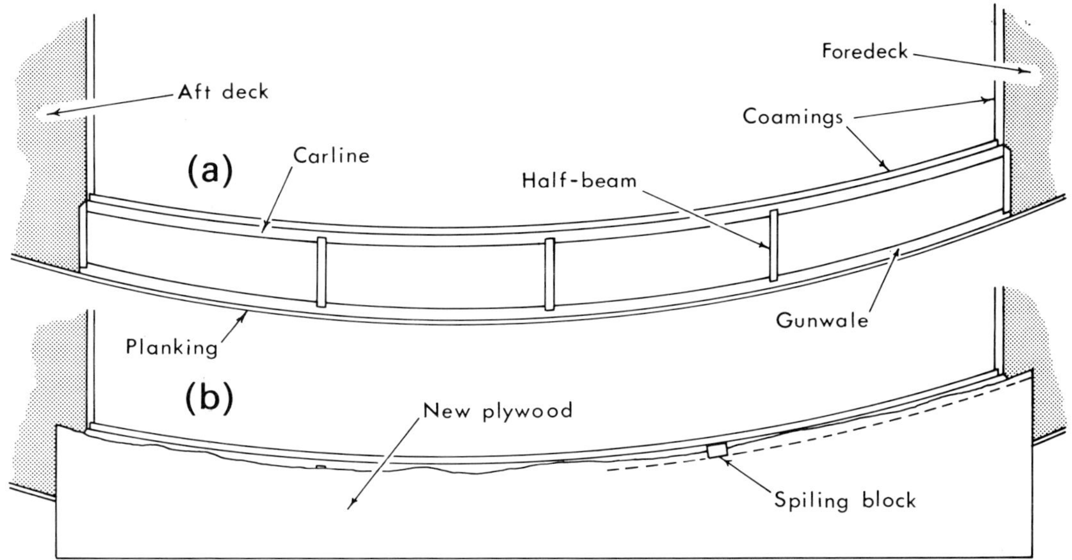

24 (a). Side deck after removal of damaged panel.
(b). New plywood panel being spiled.

steam lance or repeated dousing with boiling water.

One great advantage of using hardboard for a template is that a rough shape – even considerably undersized – will suffice. A few measurements beyond its edges are all you then need to enable a true shape to be transferred to the plywood. The best procedure is to run a spiling block and pencil all around, saw and plane the hardboard dead to these lines, then reverse the pencil position on the spiling block and run it around the planed edges while the hardboard is held firmly on top of the plywood.

Similar methods are used to transfer queer shapes (such as bulkheads) in many types of boat repair. Refer to Chapter 9 for the repair of holes in plywood panels.

Drop Keels

Dagger, drop, or swing keel plates and boards, also rudder blades, receive a lot of damage in certain sailing grounds, especially when crews are inexperienced. Hinged boards of solid or laminated mahogany sometimes rot inside the case, but more often they break clean off near the keel or get their lower or leading edges chipped or split.

New wood scarfed into the edges (see fig. 25) can make a sound and lasting job. When a really big inlaid repair becomes necessary, the board is bound to be weakened, so it often pays to fabricate an entirely new board and keep the old one as a spare.

Metal plates and rudder blades more frequently get buckled – especially those of light alloy. Few amateurs are equipped to tackle such stiff plate (see also Chapter 9) and the assistance of a shipyard or engineering works is called for. A straightened plate is never quite the same again so if the dinghy is of a type still in production, a new plate or blade from the makers could prove cheaper than having one made locally. Welded repairs to steel plates involve first grinding away the zinc coating locally. Re-galvanizing is then necessary to produce a sound job. With the correct equipment, welding light alloy presents few problems.

Pivot bolts wear rapidly on some craft left afloat at moorings and early renewal is always a good idea.

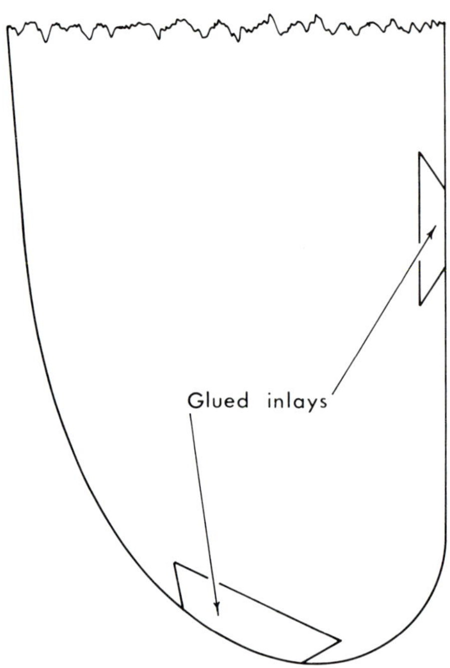

25. Drop keel edge repairs.

Other Dinghy Repairs

The treatment of slow leaks, large and small holes, also framing repairs – for wood, metal and plastics – described in the ensuing chapters, applies to dinghies as well as to larger craft.

When dealing with certain canoes, folding dinghies and inflatables, varying repair techniques apply, but fortunately most makers can supply instruction sheets, advice, and the correct materials, to ensure that the sensible amateur is able to eliminate many sorts of defect safely and neatly.

seven

Framing Faults

Most of the repair work described in the last chapter concerns typical external problems which most amateurs could tackle. Reinforcing weak interior framing (of timber or metal) by fixing new pieces alongside old demands about the same amount of expertise.

Making invisible repairs to such members, however, entails stripping away all the old material and inserting new. This can be quite a test of skill. Be consoled by the fact that difficult internal repairs undertaken by enthusiastic amateurs frequently turn out neater and stronger than similar boatyard work. The professional is quicker – having done such tasks before – but in most cases he must forgo some precision to make the job pay.

Rot Treatment

Sea-water is a mild antiseptic, inhibiting the growth of rot fungi and other organisms which cause timber to decompose. Rot most commonly occurs where rainwater seeps through leaky decks and lingers in badly ventilated places. Any part below the waterline is also prone to rot in craft based permanently on fresh water lakes and rivers.

If rot can be detected in the early stages, serious trouble can often be averted by fungicide treatment, by eliminating the cause, or by a combination of the two. Interior fitments tend to hide things and the musty smell emitted by rot is often considered to be the natural odour of a badly ventilated boat.

Discoloured, blistered, or cracked paintwork may be the first sign of trouble. Boat surveyors test for rot by tapping with a hammer – a dull tone means softness. A thin spike only finds trouble when the rot comes close to the surface being probed; see Chapter 4.

Chemicals

Rot treatment may start at any stage of deterioration but structural weakness usually demands renewal of the affected parts. Getting the popular anti-rot liquid chemicals (such as Git-rot or Cuprinol) to soak right into suspect timber is not easy. It must be done under bone dry conditions by repeated dousing, or by means of wicks – one end of each poked into a hole bored into the affected timber, the other end submerged in a pot of the chemical.

For permanently damp conditions, porous bags of crystals are available. These are packed around the faulty member, held in place by means of nails and battens, wedges, or struts. The chemical is gradually dissolved by the inherent dampness, and thus saturates the timber.

The oil-soaked bilges of motor fishing vessels and workboats never seem to rot. Kerosene (paraffin oil) has long been used as a preservative for soft old stems, frames and deadwoods, but this does not exactly improve the fire risk on a boat. Unlike diesel fuel, the smell of kerosene disappears after a few days.

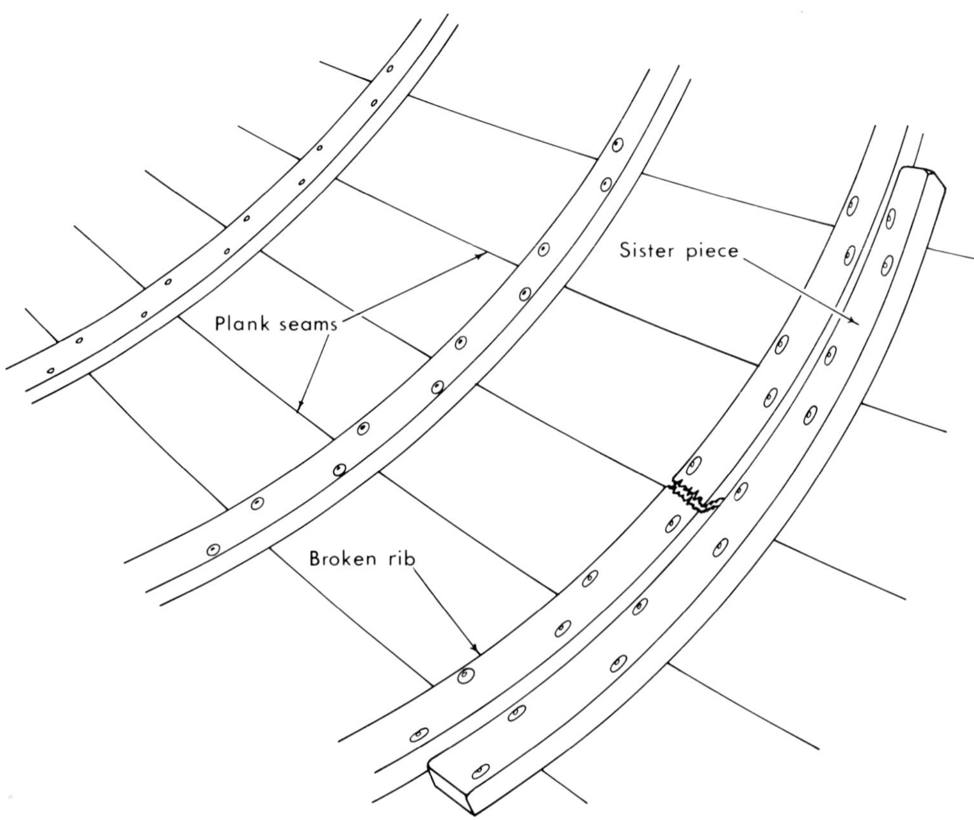

26. Sistering a fractured bent timber.

Woods like teak are virtually immune from rot, but other species should be well soaked with a good marine preservative (such as PCP or Celcure) when new wood is used for a repair job.

Rot does not appear in boats of plastics or light alloy, but its counterpart in steel craft is rust. This can be held at bay readily by chipping, dosing with phosphoric acid and maintaining a substantial coating of steelwork paint at all times, as described in Chapter 5. Should this treatment be left too late, replacement of metal will become necessary – see later in this chapter and in Chapter 9.

Steamed Timbers

The somewhat frail bent ribs found in clinker boats (also in most small round bilge wooden sailing craft and in some powerboats) need more repair work than any other item of internal framing.

The most common fault is a clean cross fracture in line with a planking seam. A single fracture can be ignored, but two in one rib, or in adjacent ribs near the same seam, need immediate attention. The most common method is to fix a sister piece alongside, see fig. 26. Shipwrights call the process *doubling up* or *sistering*.

The new pieces should be of similar width and thickness to the original timbers, but when they must be fed behind a stringer or riser, reduced thickness is often unavoidable. Do not create a water trap between the new and old parts – either insert bedding compound or leave a space to permit future painting. Exposed ends and corners should be chamfered for neatness.

Straight-grained oak is usually chosen, but American elm (Canadian rock elm) is even better, while hickory, agba, hackmatack, or ash are possible alternatives.

A sister piece needs to cover about four plank widths to function properly. Some will bend into place cold, while others can be given sufficient shape with plane and spokeshave. Otherwise, preform the necessary shape by cramping the new piece to a rough jig (plate 15) after steaming or submerging in boiling water for about one hour per inch (25 mm) of thickness. Allow the piece to dry out before fixing so that the back may be painted.

Re-ribbing

Where serious cracking has occurred on many ribs, or where doubling-up would look unsightly, two alternatives offer themselves. Neat looking extra ribs may be added exactly mid-way between each original timber throughout the boat. These normally extend from the keel just past the turn of the bilge and look as if they were always there. In the second method, all existing ribs are removed completely (two or three at a time) and exact replicas are inserted using the original plank nail holes to fasten them.

Needless to say, just a few faulty timbers may be cut out and replaced to avoid the unsightliness of sisters. When all need renewing it sometimes pays to remove the stringers or risers, thus obviating the need to strip off gunwale cappings or decking. Many craft have timbers in one length from gunwale to gunwale, but to simplify fitting replacements, make each one of a separate piece port and starboard, meeting side by side on top of the hog or keel.

Chine Frames

Most chine (Vee-bottom) boats, whether with plywood, carvel, strip, or multi-skin planking, have varying numbers of sawn frames. Being stiff, such frames need a bad collision to break them, but as they are normally cut from straight-grained stock, cracking along the grain (fig. 27*a*) is quite common where a hull has a lot of curvature or flare. Similar splitting can also occur along the line where the points of all the plank fastenings lie.

A short crack in a single frame is not normally detrimental. One or two long screws into the inboard edge will add some strength and prevent the crack from lengthening, while stopping and paint will improve the appearance. A sister piece bedded in mastic on the flat side of a frame (and screwed or through-bolted to it) makes a strong job, especially when additional plank fastenings are driven into it. However, it does not take much longer to renew a whole frame leg, either from keel to chine, or from chine to gunwale.

Crook Frames

The curved sawn frames used in round bilge yachts of heavy construction – sometimes with two or three bent ribs between each sawn one – present different problems from their chine counterparts. Although this type of frame is often built up nowadays by laminating thin strips with resin glue, in the older craft which most frequently need repair, sawn frames are made from natural *crooks* (sweeps) cut from the big limbs and curved trunks of oak trees. A certain amount of sapwood and other defects creep into sawn frames, so shakes and rot are not uncommon.

Although laminated frames are made in one piece from keel to shelf, it proves almost impossible to wriggle a new one into position without stripping out other parts. A scarfed repair as in fig. 27*b* is therefore the most common procedure.

Crook frames are often simpler to deal with, as the originals are normally made with one or more butt joints strapped across with doubling pieces. The bigger ones are made as double frames all the way with joints staggered in each half.

With careful use of mallet and chisel, one crook only may be broken out or scarfed without touching its neighbour.

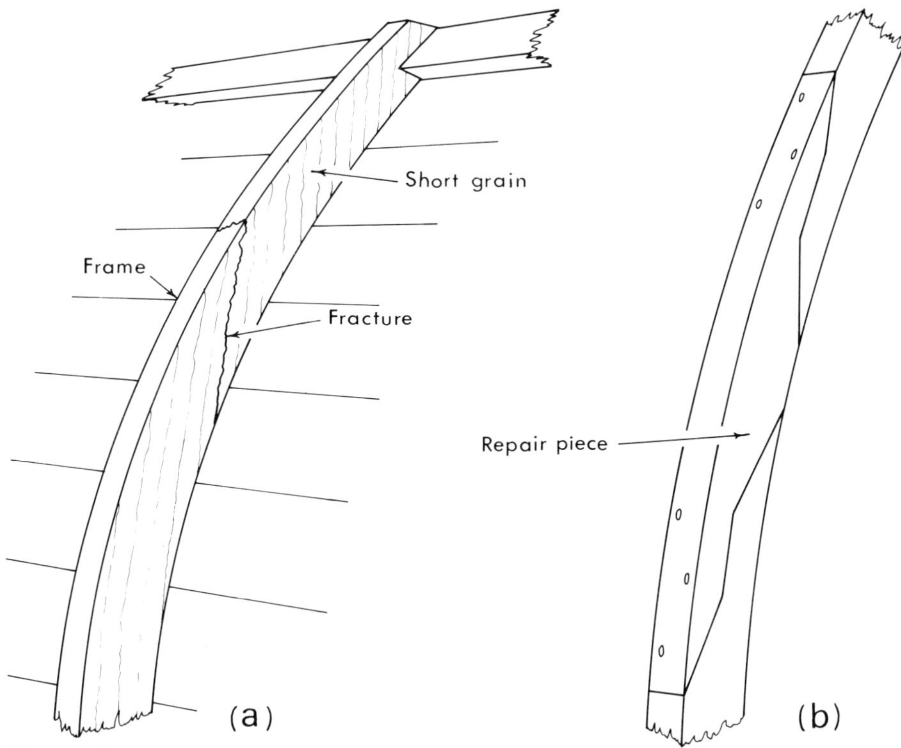

27. Sawn frame repair.

Seasoned natural crooks of oak are sometimes difficult to obtain. Where a damaged frame has little curvature, any good straight-grained hardwood will do. Otherwise, laminate the necessary new piece from thin strips of hardwood glued together in leaf-spring formation. The same idea can be used for replacing steamed ribs or making sister pieces – laminating direct into the boat or over a jig on the bench.

Copper Fastenings

In wooden craft of all ages, riveted copper nails are commonly found fastening the skin to bent or sawn frames. Withdrawing old copper rivets which are not buckled is normally a simple matter of clearing any stopping from the head, filing off the riveted part, then driving out the nail with a punch and hammer until the head protrudes sufficiently to enable pincers or a claw extractor to be applied. New ones are quick and easy to drive once the knack is acquired. They pull a joint together tightly and further riveting is possible at a later date should the timber shrink.

Pilot holes are essential. The drill size is readily found by a few trials, allowing a tight driving fit without fear of buckling.

Using Roves

The nail is driven through while a heavy hammer (or a chunk of thick steel bar called a *dolly*) is held close to the pilot hole inside the boat. The dolly is then held over the nail head while a cone-shaped washer called a *rove* or *burr* is forced over the protruding nail point with a hollow rove punch. Such a punch is easily made from a sawn-off steel or brass bolt with a hole of appropriate size drilled centrally into the end.

Wire cutters are used to crop off the surplus nail about one nail thickness proud of the rove. Using a ball pane hammer with light rapid strokes (while an assistant holds the dolly over the nail head) the protruding bit of nail is riveted neatly over the rove. Finish off with a few blows using the flat tip of the hammer, but do no over-rivet causing the rove to flatten.

Although domed rosehead copper nails are available, countersunk ones are more often used. With clinker planking, the flat tops of these nails are usually set flush with the surface. For carvel, plywood, strip and double diagonal planking, the heads are normally sunk below the surface and stopped or dowelled over.

Note that in this latter case, the dolly used must have a protruding stud (the same diameter as the nail head) fitted into it or welded on. To avoid having to make this tool when only a few nails are to be driven, a nail punch of correct tip size may be held between the dolly (or a heavy hammer) and the nail head.

Turned Nails

Copper boat nails are occasionally used without roves. Especially where the work is hidden from view or painted, the points may be bent back into the timber and hammered down flush. With a dolly over this end while the head is treated with hammer and punch, a good tight joint results. Turned copper nails are mainly used nowadays for joining sheets of plywood together when a butt strap of similar plywood is fitted inside.

Being so smooth, copper nails do not work very well when driven straight into thick timber like common wire nails. On the few oc-

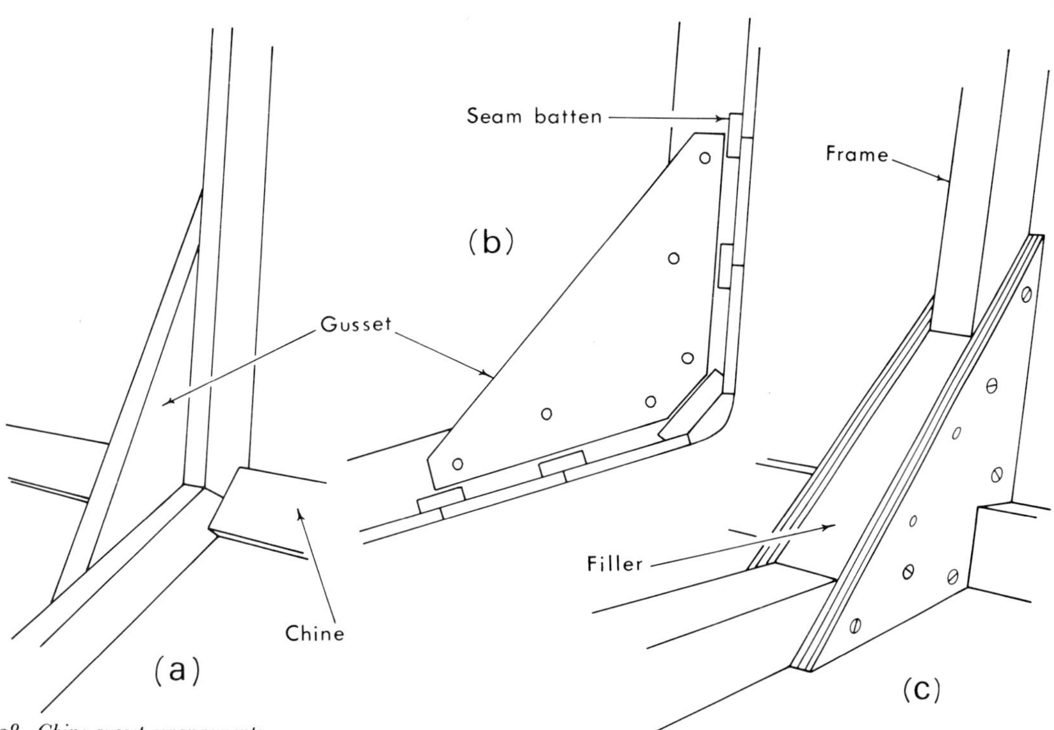

28. Chine gusset arrangements.

casions when this is necessary, holding power may be improved considerably by twisting each nail a full turn from head to point.

The risk of buckling when driving into a tight hole is lessened by first rubbing the nail with candle wax or tallow.

Chine Gussets

The separate parts of chine boat sawn frames are usually jointed together at keel and chine – also sometimes at the gunwale where a deck beam joins a frame – by gussets. These are triangular plates of straight-grained timber or plywood (fig. 28*a*) glued and screwed (or nailed) to one or both sides of each joint. Gussets fail most frequently by the cracking of unsuitable timber or rotting caused by moisture trapping.

Removing glued gussets is usually a messy hammer and chisel job, but making and fitting replacements is a simple matter calling for cardboard templates, glue and screws. If the original gussets had notches to receive seam battens (or the thin stringers used in some double diagonal boats) do not attempt to reproduce these in a new gusset – just butt it against the battens as shown in fig. 28*b*.

Retain any filler chocks between pairs of gussets (fig. 28*c*) as these help to prevent moisture-trapping and to protect thin gussets from damage. Defective old fillers should therefore be replaced, set in mastic unless made accurate enough for gluing.

Twin gussets on large craft are sometimes through-bolted. Avoid boring fresh holes when renewing such gussets, as this could weaken the frame. Offer up each gusset in turn and mark the hole positions through the frame by waggling a pencil, tapping an improvised centre punch made from a bolt, or making dabs of paint with a long thin brush. A system of numbering or lettering is advisable to make sure that several new gussets are correctly located.

Floors

Gussets across the keel or hog are called *floors* by shipwrights. In big round bilge craft they are laminated or made from grown crooks. Such floors are sometimes fitted between the bent timbers of motor launches and sailing craft to add strength, to support the engine bearers, to receive ballast keel bolts, or to strengthen a drop keel trunk.

One cracked floor matters little, but two weak ones near each other must be replaced. Any soft or rotting timber demands immediate elimination to avoid the danger of spreading to adjacent parts.

With clinker planking, floors should be joggled over the lands, as in fig. 29*a*. An accurate hardboard template is almost essential to prevent mistakes when shaping such a floor. Especially towards the ends of a boat, the mating surfaces of floors have bevel, one face being smaller than the other. For wide floors it sometimes pays to make a separate template for each face.

In small craft, floors may be screwed to the planking, but very secure keel fixture is necessary, usually bolts or coach screws, sometimes combined with metal angle brackets.

Many wooden craft have floors of galvanized steel or bronze. When damaged, these should be removed for welding, brazing, or making new. Such repairs are sometimes possible in situ, care being taken to protect adjacent timber from heat damage by wetting it and arranging covers of sheet metal and asbestos mats.

Floors built into metal hulls may be repaired in similar fashion. To strengthen the keel and bottom of a fibreglass boat, fitting additional timber floors with closely spaced stainless steel woodscrew fastenings through the skin is often more satisfactory than attempting to bond new glass mouldings to the existing laminate.

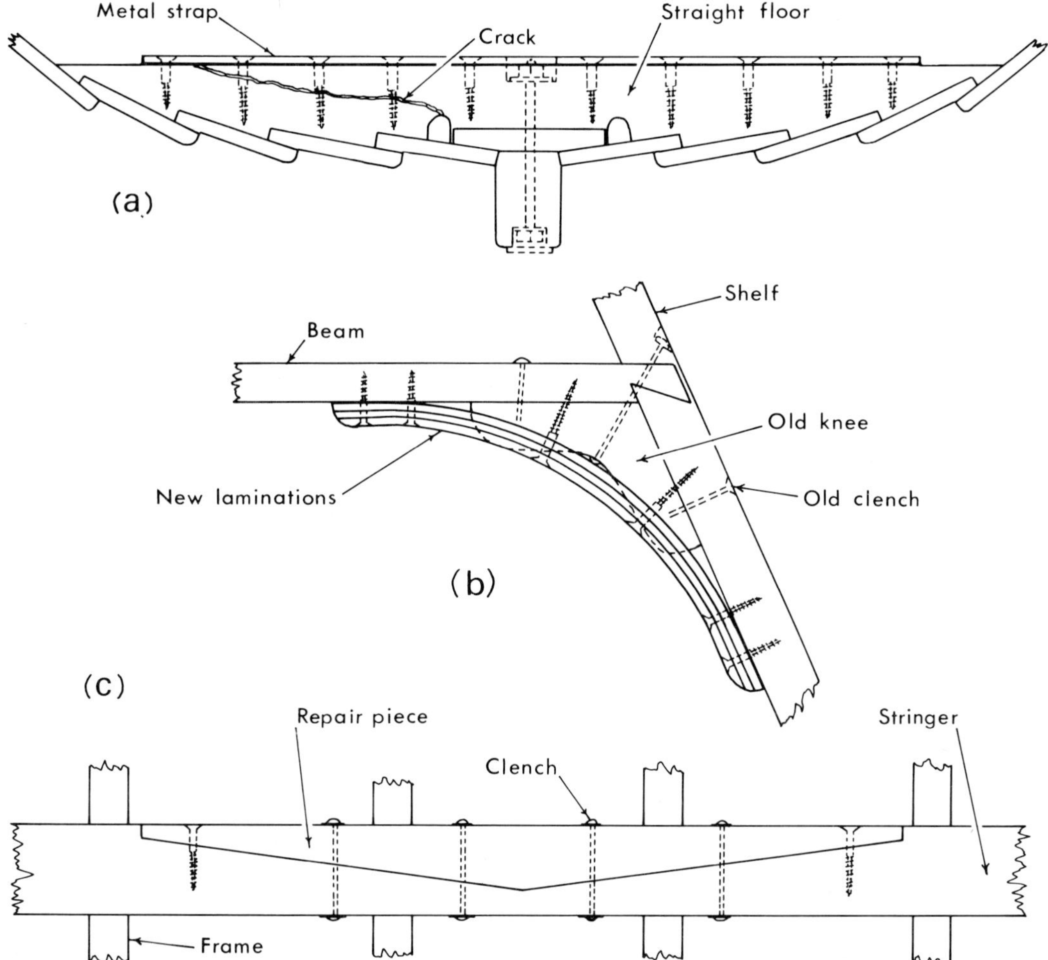

29. Some floor, knee and stringer repairs.

Small Knees

Although similar to floors, the small knees used for stiffening deck beam ends, securing dinghy thwarts and gunwale ends at bow and stern are easier to repair. Traditional ones are cut from natural crooks of oak, apple or cherry, but offcuts of plywood glued together (to build up the required thickness) work well.

Reinforcing a cracked knee is often possible by cutting back and adding two or three laminated strips as in fig. 29*b*. A bent metal strap will sometimes serve the purpose where its appearance is not detrimental.

Knees require long through-fastenings; screws are almost useless. The favourite method is to bore a push fit for round copper bar and rivet this over a large rove or copper washer at each end. Bolts are ideal if properly recessed and stopped over. Sometimes there is no way to get a standard bolt with head into place, but a length of bar threaded for a nut at each end solves this problem. Continuous threaded bar called *studding* is often available from chandlery stores and is widely used for jobs such as this.

Stringers

Longitudinal framing members are not usually so prone to rot, fractures and splits, as ribs and knees. Accessibility to stringers is generally not too bad over the short length involved in most repair work. Although much patient whittling with mallet and chisel is unavoidable, a piece may be scarfed in neatly as shown in fig. 29*c*, without undue loss of strength. Doubling up with a sister piece is quicker, but looks unsightly and usually means disturbing the exterior of the planking to drive new fastenings. Instead of sistering alongside, one can sometimes plant a strengthener flat on the inboard face, bolted through the old stringer by fitting the nuts between stringer and planking. The new piece can usually be planed to fit the curvature without recourse to steaming or lamination.

An alternative method is to fit galvanized steel *fishplates*, either a single flat plate bolted to the face, or angle section above and below.

The short stringers receiving the bolts from bilge keels are so important that complete renewal of defective ones is usually wise. A loose bilge keel could sink the ship!

Beam Shelf

Although the shelf on a decked boat is similar in shape to a stringer and takes the place of the inwale of a dinghy (see Chapter 6), a shelf supports the beam ends with dovetailed joints and extensive repair work is fraught with difficulty.

Damage below beam level is tackled in similar manner to the stringer work described above, but rot on top of a shelf is common due to deck leaks. Breakage due to a serious collision usually means that the deck is damaged

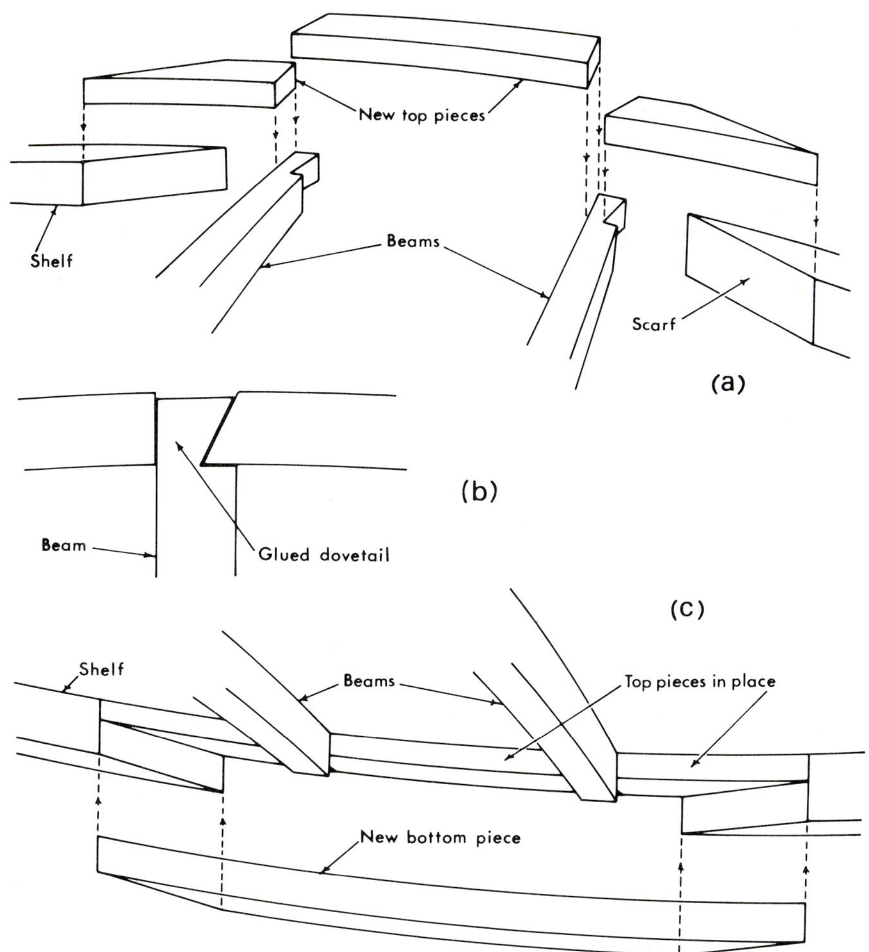

30. Replacing a length of decayed beam shelf.

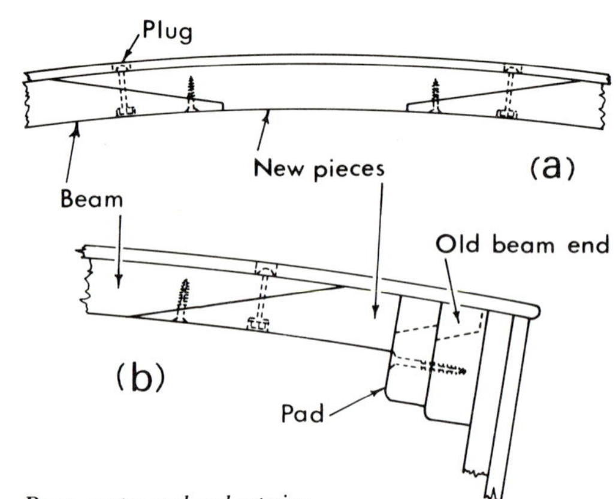

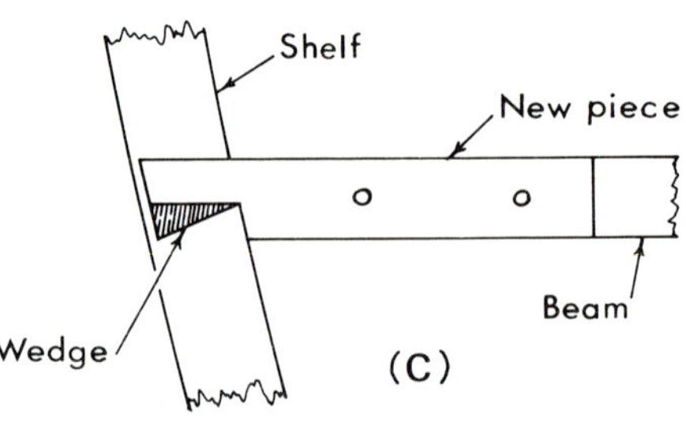

31. Beam centre and end repairs.

also, but stripping part of the deck simplifies work on the shelf considerably. In such cases a little help from a shipwright is generally prudent.

If the beam ends are sound, a strong amateur repair is possible by using partial lamination. The shelf is cut away and scarfs formed on the sound timber at each end. Separate pieces are made to fit between the beams, see fig. 30*a*, any curvature being planed out of extra thick timber. Templates are advisable to ensure good glued joints against the dovetails as in fig. 30*b*.

Next, a baulk is prepared to make the remainder of the shelf in one piece (fig. 30*c*) scarfed to the existing parts and curved as necessary by steaming or planing.

After offering up with all bolts and screws fitted temporarily, prepare wedges from the deck and struts from the stringer to provide effective cramping on final assembly with glue.

A shipwright might save time by making the whole thing in one piece to slide up from beneath, dovetails and scarfs fitting perfectly in unison. Amateurs generally prefer the more prolonged but foolproof method!

Lodging and hanging knees can form major obstacles to this type of repair. Metal knees are usually easy to remove intact, but wooden ones may have to be destroyed to get the fastenings out.

Beams

Rot is the main source of defects in deck beams, followed by cracking and warping of the original timber, plus possible collision damage.

Trouble near the middle is usually dealt with by scarfing in a new piece as seen in fig. 31*a*. If free from rot, a fishplate bolted to one or both sides could solve a cracking problem, after first jacking up the beam to remove any sag. With all beam repairs, the proximity of carlines or bulkheads can add greatly to the amount of work involved.

When a beam end fails, there is no need to renew the whole beam. Scarf in a short piece making a new dovetail set into a pad planted on the inner face of the shelf as seen in fig. 31*b*. If this method would look ugly, glue a small wedge into the original shelf dovetail (fig. 31*c*) to transform it into a parallel sided notch for the new beam piece to slide and glue into.

Laminating in situ is often a good method when replacing either a whole beam or a small piece. However, using vertical strips (instead of horizontal ones leaf-spring fashion) has advantages. The difficult task of cleaning up the sides is then eliminated and each strip can be bowed sideways to engage the shelf notches.

Metalwork

Boats constructed of ferrocement and plastics do not have separate framing comparable to craft of wood and metal. They do not rot and, when any internal frame-like reinforcements are damaged, the whole skin is also damaged and repairs to both are carried out in one operation as described in Chapter 9.

Steel corrosion compares with the rotting of timber, but light alloy boats are generally only damaged by collision or grounding. In general, repairs to metal framing are easier and cheaper than comparable wood renewal, but few amateurs are equipped to tackle it, or have the necessary expertise or confidence.

Thin steel is difficult to weld reliably when even slightly rusty. Therefore, any cutting out or reinforcing of framework should be extended well beyond any such weakness. The temptation to repair with thicker metal than the original should be avoided, as both gas and electric welding works best if both parts are of equal gauge. Scarfing is not normally necessary as any doubtful joint can be reinforced with a doubling plate.

Galvanizing or zinc spraying is anathema to welding (see Chapter 9) but its removal by grinding is possible except in awkward corners. Instead, it often pays to adopt riveted or bolted repair work – a sound idea much revered by amateur metal-bashers!

Frames sometimes rust away completely in the bilges, even when originally galvanized. If riveted to the skin plating, neat replacement of the rusted sections is possible. If welded, it could pay to double up with new frames and floors mid-way between the old ones.

Frames buckled in a collision can sometimes be jacked into shape. More often, they must be cut right out and new metal welded or riveted in. A curved frame needs a template before rolling or before bending hot. Any engineering shop will apply the necessary curvature to angle or T-section metal, but remember to leave them a respectable amount of spare length, as few bending machines can operate right up to the bitter end.

When the bilges of new steel craft are concreted (perhaps encasing the ballast) this acts as an excellent preservative. When concrete has been placed to hide trouble, it can prove highly undesirable. Breaking out small quantities is not difficult, but removing all the concrete generally necessitates stripping off skin plating in the vicinity.

For a small craft of light steel or alloy construction, framing repairs are greatly simplified by the use of pop (self-closing) rivets. These are available in aluminium, monel (cupro-nickel) and stainless steel from tool stores, together with the guns to close them. They are particularly useful for box sections having no interior access.

With all metalwork the removal of sharp corners and edges is most important for the sake of appearances, safety and good paint adherence.

Backbone

The most troublesome of all framing repairs in any material concerns stems, keels and sternposts. Although the repair of these parts is generally best left to the professionals, amateurs do tackle them satisfactorily when given sufficient time. Many an old boat has been scuttled rather than face such major problems!

A stem or sternpost is replaceable without disturbing the planking. The simplest procedure is to laminate the new parts from both inside and outside the boat, thus forming a new rabbet around the planking ends.

Cutting out the old parts without causing unnecessary damage takes a long time. Few fastenings come out readily the way they went in, so one must resort to cutting them off, boring them out and chopping around them.

To replace a keel, it pays to remove the adjacent (garboard) strakes of planking first. Any hog piece is then completely exposed as well as the keel proper. However, with plywood, cold moulded, or double diagonal planking, you cannot expose the keel without cutting away a large expanse of the skin if loss of strength is to be avoided.

Some old launches have very light keel structures. When renewal is necessary, increasing the scantlings is comparatively easy provided new garboards are fitted or the originals re-shaped.

Bilge keels can also be made more substantial. When the tips of wooden ones become damaged or rotted, new timber can generally be let in (and the bolts withdrawn one by one if necessary) without disturbing the hull joint. Steel bilge keels are simply bolted to the hull via flanges, but complete removal is advisable to ensure that a reliable downhand welding job is done in the workshop. Bolting

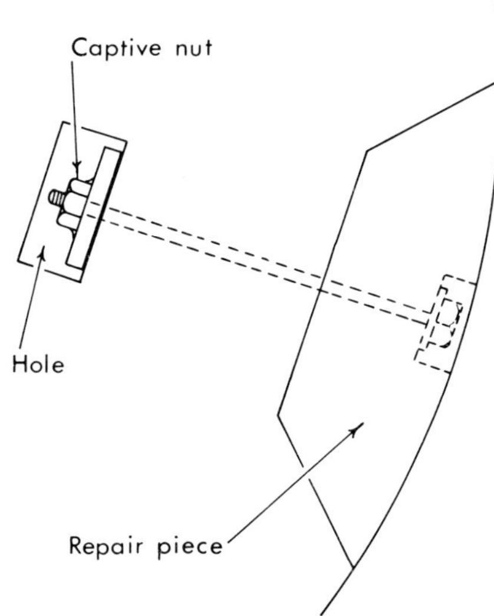

32. Using a captive nut for rudder blade repair.

metal plates to timber bilge keels is a useful device sometimes.

Bolts with captive nuts (fig. 32) are useful to fix a repair piece into a massive timber part (such as a bilge keel, rudder, or engine bearer) or to reinforce an existing joint. With the nut welded to a square washer, no spanner is required at that end, eliminating the need for a big aperture. By tapping a thread diametrically through a chunk of big bar to make the nut, this can be housed in a simple bored hole. A little ingenuity is necessary to position such a hole precisely. Captive nut apertures should be neatly sealed with glued timber plugs on completion.

eight

Leaks

You still need at least one good bilge pump on a boat of fibreglass, metal, or ferro-cement, for although such vessels often have bottle-tight hulls in comparison with wood construction, bilges rarely stay dry due to rainwater, spray, condensation, faulty stern glands and accidental spillages. Unless in need of repair, a plywood, cold moulded, or strip planked hull should be dead tight. Carvel and clinker boats leak notoriously after long spells hauled out, but they should take up completely after a few days afloat. Bad cases need pumping night and day for a while. Those with softwood planking take up more readily than those of teak, mahogany, oak and similar hardwoods.

Locating Leaks

If a lot of water still gets into a wooden boat after allowing a fair time to take up, a search for leaky seams, keel bolts, or skin fittings is called for. This could take a long time to do thoroughly, removing sole or bottom boards systematically from bow to stern, together with bunk boards, locker doors and perhaps ballast. A powerful lamp or inspection light helps. A keen ear can often detect the worst leaks.

Make notes of the exact locations, using measurements from salient points and counting the number of seams upwards from the keel. When the boat is hauled out again, you want to know exactly where all leaks are located to save slipping several times. Seam caulking may look sound from the outside, but the fault will soon be revealed when soggy stopping and caulking is raked out. Do not forget to replace the antifouling paint on completion of such emergency works!

To save slipping, seam leakage or a puncture is often easy to master temporarily by clapping a pad of wood coated with mastic over the spot and cramping it there with struts, wedges, or battens screwed to adjacent frames. A sheet of thick soft rubber sometimes works better than mastic. Even if not completely successful, such a repair may keep the leakage within reasonable bounds indefinitely. Do not attempt to use resin and glass mat on a moist surface.

After hauling out at laying-up time, doubtful seams will stay damp after the remainder of the hull has dried off. That is the best time to take notes – particularly to locate leaks in old double diagonal planking, as the point of emission inside the hull does not necessarily correspond with the external entry point.

Wooden boats with fin keels and external ballast often leak when sailing due to twisting of the keel. If you have shallow bilges they need regular pumping under way to prevent water sloshing into lockers and on to bunks.

Leaky Decks

Drips from the deckhead – invariably right over a bunk – are by no means exclusive to old wooden boats. Ferrocement decks are the

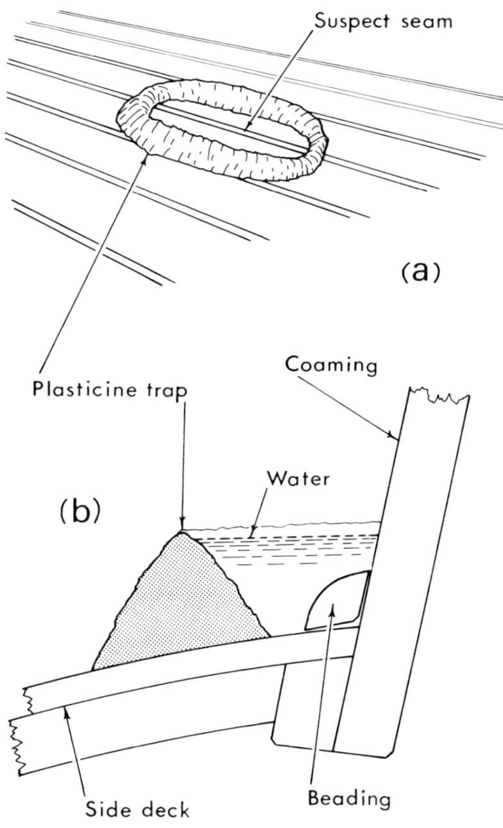

33. Simulating deck leaks.

least prone to leakage while caulked seam laid decks are the worst. The bolts securing grab rails and other fittings weep notoriously even through fibreglass decks. Plywood tends to flex and get leaky joints. Canvas coverings crack and tear, while PVC coverings shrink and curl at the edges. Fibreglass sheathing sometimes parts from its plywood foundation.

Most types of deck leak are troublesome to locate, often emerging far from the point of entry. Furthermore, simulating wet conditions is difficult to achieve at the appropriate time.

Provided there is some clue to the likeliest entry point, a small embankment of plasticine can be formed around it (fig. 33*a*) and the resulting basin filled with water. A delay of as much as one hour could occur before any internal drips show. If the deck slope is not too great, one can sometimes reduce the size of the basin in stages until the trouble spot is isolated.

A somewhat better method for detecting slow deck seam leaks is to cut the bottom off an old plastic bowl, seal this to the deck with plasticine and put a weight on top to hold it firm. The extra depth of water when this is filled accelerates the rate of leakage.

Similar experiments are equally beneficial at coaming-to-deck joints (fig. 33*b*) and their patient use could save much work stripping off and replacing beadings.

Hatches

Leakage of spray through a sliding companion hatch is not normally a disaster. Rain should not get in, but as such hatches stay partially or fully open for much of the time, spray is sure to get below occasionally.

A fore-hatch is not often situated directly over a berth, but being vulnerable to spray it needs to be tight, preferably of the double coaming type with weep holes through the outer barrier.

Adding rubber sealing strips (as used for house windows) sometimes checks such leaks effectively. When the coamings are too low (allowing the hatch to get submerged momentarily in green seas) it pays to make a new lid of increased size, raise the original coamings and fit additional double coamings with weep holes all around, rather like the *dorade* type sprayproof ventilator shown in fig. 50.

A tight-fitting canvas cover keeps out a lot of unwanted moisture in bad weather. The varnish-preserving qualities of hatch covers were mentioned in Chapter 2.

Skylights

In the same vein as above, a skylight cover works fine provided it has transparent panels on top. A drawstring in the lower hem is generally more satisfactory than shockcord. A correctly made traditional teak skylight should not leak unless submerged, but when there is trouble, make sure that the weep holes from the gutters are clear of debris.

Cheap skylights have butt hinges and thin coamings with no gutters. Leakage can usually be reduced by fitting continuous (piano) hinges, rubber sealing strips and internal latches which batten the lid down tightly.

Caulking

Before you buy cotton and marine glue to re-caulk deck seams (plate 16) make sure they really are traditional caulked seams. Modern yachts with teak decks generally have a plywood backing. The seams do not need caulking and are filled instead with epoxy resin or synthetic rubber composition.

Complete traditional deck re-caulking takes a great deal of time but, having a horizontal surface, work is far less laborious than when caulking vertical topside or overhead bottom seams. You should be able to borrow the necessary tools from a friendly boatyard; few chandlery stores stock them nowadays.

Hooks for raking out old cotton or oakum from the V-shaped seams are readily made on the tang ends of old flat files. If they refuse to bend cold, heat the metal to dull redness.

Ask an old shipwright for a few tips about driving cotton correctly – you will soon get the knack. Experiments are always necessary first to gauge how many strands of cotton you need to fill each seam of a certain width. They are rarely all alike.

Having ascertained this, make up a few bundles of twisted strands, ready to drive into the seams with a caulking iron. Pulling from the middle of a ball of caulking cotton you will normally find a cord of eight loose strands.

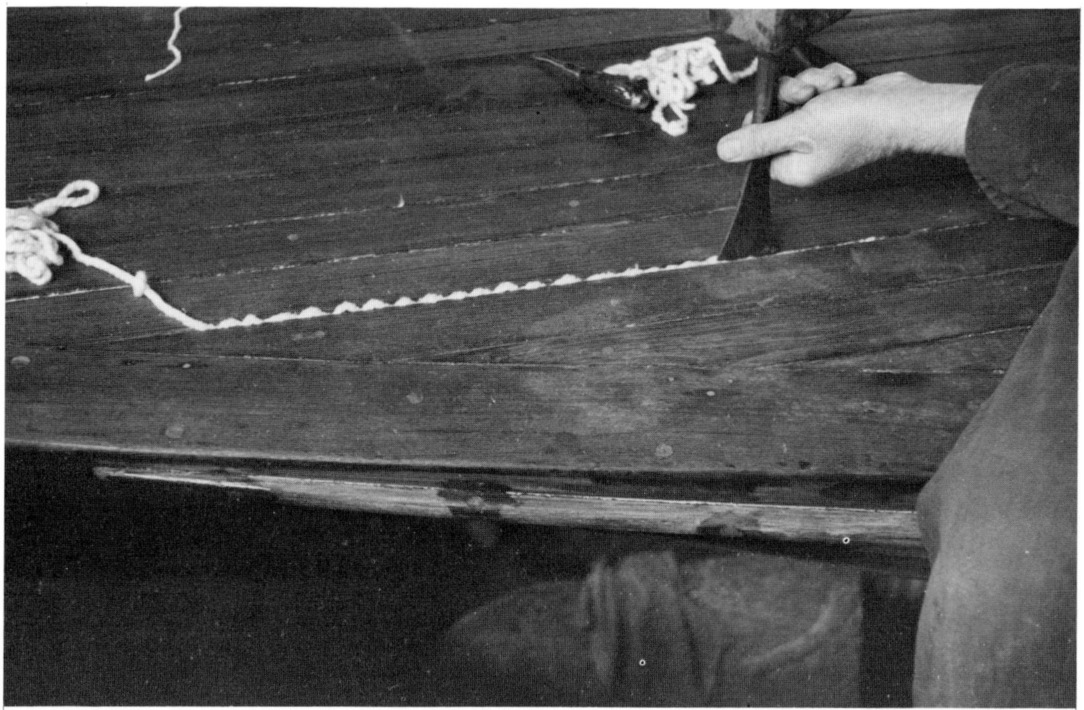

Plate 16. Driving twisted cotton into a deck seam.

Cut off about 20 ft (6 m) and part from this the two or three or more single strands needed for a particular seam. Make fast one end, then tie the other end to a bent-over nail held in the chuck of an electric drill, or hand drill. Keep winding until the whole length turns into a rope and starts to pull away from you. Disconnect both ends and flake it all down into an empty cardboard box.

While feeding cotton into a seam, pinch it down firmly every 2 in (50 mm) using the corner of a narrow-tipped caulking iron and a light mallet. Now go back over the seam, first using the narrow iron, then repeating the process with a thicker iron which just fits freely into the seam, this time using a heavy mallet. On completion, the cotton should lie about one seam width below the surface. Inspect the deckhead periodically to ensure that you are not driving so hard that loops of cotton are protruding below decks!

To join lengths of cotton, intermesh the strands for about 2 in (50 mm) and twist tightly. To join new cotton to old, just bed the ends in plenty of soft mastic sealer squeezed from a tube or gun with nozzle. Solid deck planking is sometimes strengthened by setting metal dowels across the seams. Dowels should lie below the caulking Vee, but check for proud ones which could damage the caulking iron. Except for this, caulking hull seams (both above and below the waterline) is almost identical to the procedure for decks.

Paying Hulls

Cotton swells when moist and seals a seam tight, but to work effectively it must be boxed in with the correct type of stopper known as *paying*. Excellent products are marketed for this job – special types for decks, topsides and bottom seams, most retaining some degree of elasticity. With help from stockists, boatyards and makers' instructions, one should have no difficulty in choosing the best brand for a certain application. A broad stripping knife (palette knife) is the most effective tool for paying, drawn across the seam at right angles, forcing the stopping down onto the cotton. Although a caulking gun will apply stopping rapidly, a palette knife obviates voids and leaves a surface which needs little subsequent finishing work.

Paying Decks

Often having no paintwork protection and suffering climatic extremes, the paying of deck seams generally needs more maintenance than that of hull seams. Traditional marine glue run hot from a ladle (plate 17) along each seam in turn, then scraped off flush when cold, is still widely used and is by far the cheapest method. A little professional guidance – perhaps the hire of a boatyard expert for a few hours in his spare time – is a good idea at the start. Choose a long spell of dry weather, as the wood must be bone dry.

Glue temperature is important. To avoid air bubbles, one slow pass is better than two rapid ones. Take care not to overflow too much, or there will be a lot of scraping to do. Professionals use scrapers made by forging the wide ends of big flat files into ground and hardened hooks. The amateur is normally happier with the interchangeable blade type of scraper sold by tool stores.

Modern teak and pine decks are generally payed with a synthetic rubber composition which may be air-hardening or catalyst activated. These materials are extremely expensive, but are ideal for amateur use and

Plate 17. The traditional ladle for pouring hot marine glue.

allow the timber to expand and contract without leakage. They are intended to fill the seam completely, eliminating the need to drive cotton. However, material costs can be lowered by driving a little cotton first. Application may be by gun or tube for the one-shot compounds and by palette knife or syringe for the catalyst types.

Mallet and chisel will break out old marine glue, but two tenon saw cuts work best for synthetic rubber. For success with new synthetic rubber, timber faces must be cleaned back to new wood and coated with a special primer. On old decks this normally entails the use of a power router, or some patient work with sandpaper glued to a timber wedge. A power belt sander is the best tool to remove surplus cured composition.

Deck Canvas

Nowadays there are many paint-on or knife-on coatings which effectively cure surface leaks in both planked and canvased decks, but to do the job properly generally necessitates removing some or all deck fittings and beadings, perhaps also the toe-rails and stanchions. While going to this trouble, one might as well lay new canvas or adopt fibreglass sheathing.

Being available in kit form with full instructions, the latter material presents few problems to the amateur – provided he can stand the smell! Canvasing is simple enough provided you can lay hands on some untreated white cotton duck. For dinghy and launch decks, thin material like calico will do. For cruising boats, thick canvas is essential for long life. The grade used for big fairground tents is about right – you may be able to obtain a good secondhand piece from a tent maker. Sailmakers rarely use cotton nowadays, but they should be able to order it.

Removing old canvas which has been stuck down is laborious. Although chemical stripper helps to remove surface paint, a blow torch is often necessary to loosen the canvas. Correctly laid canvas pulls away easily without needing

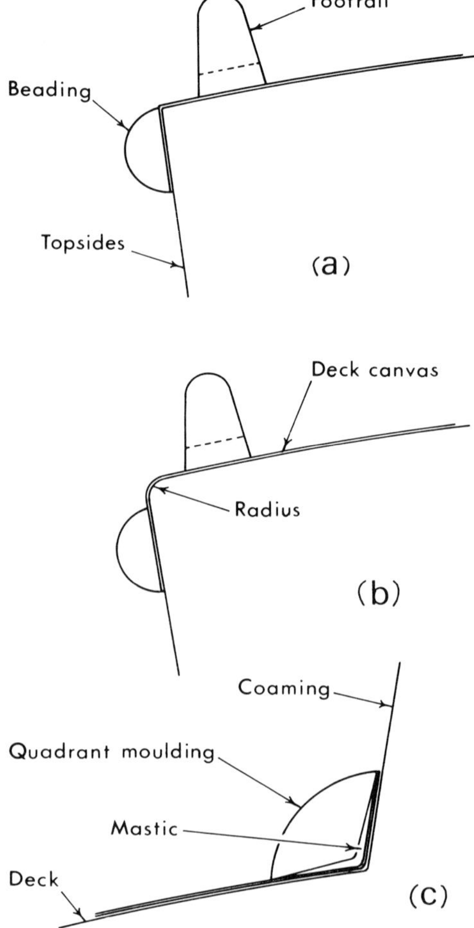

34. Improvements to deck corner edges.

stripper or heat. The bare deck must be devoid of all lumps, ridges and upstanding nails to receive new canvas or fibreglass. If earlier efforts were finished as in fig. 34*a* at the gunwales, round off the edge as in fig. 34*b* and canvas life could well be doubled at this vulnerable place.

It pays to renew all beadings unless the old ones were screwed on and are easy to remove intact. Quadrant mouldings must be planed to the correct back angle as shown in fig. 34*c* or leakage is sure to occur. Bed all beadings in mastic and use screws in preference to nails – spaced not more than eight beading thicknesses apart.

To save time, get all main deck canvas double stitched (as for a sail) into one enormous coat. If this is not possible, lay the canvas in separate panels starting where the deck dips to its lowest point and working each way from there. Turn under each exposed seam (unless a selvage) and fix with closely spaced copper tacks.

In the old days, canvas was always laid in fresh paint and then wetted to shrink it. This eventually caused cracking when the deck planking contracted in hot weather. Such cracking is not likely to occur over a plywood deck, and glue or wet paint is still necessary on hollow parts of a deck where the canvas would otherwise lift clear of the planking when stretched. Place battens and weights on top until the adhesive or paint hardens.

Start towards the centreline, stretching the cloth fore and aft, driving a few tacks and fitting beadings where necessary. Then stretch athwartships and tack along the topsides prior to clamping permanently with the half-round strips.

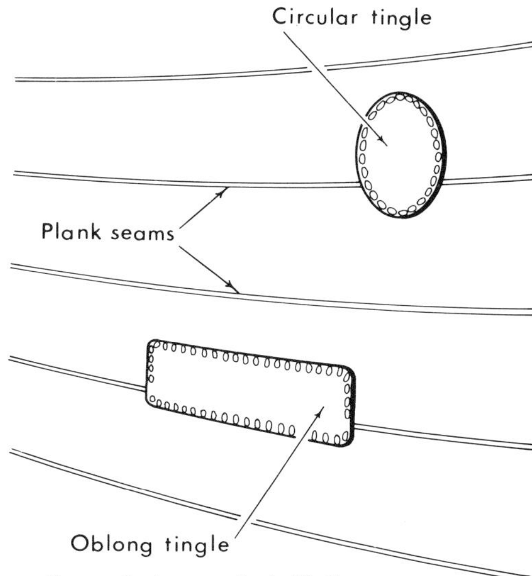

35. Copper tingles on a planked hull.

Next, cut off all surplus canvas, apply three coats of thin white lead paint, bed all deck fittings in mastic, paint the canvas with anti-slip deck paint, plug all screw holes, then paint or varnish the beadings.

Tingles

The patches fitted to a boat's bottom planking (fig. 35) to prevent small leaks are called *tingles*. In some cases they remain for over ten years quite happily before structural repairs eliminate them.

Tingles are easier to make (and look neater) if square or oblong rather than circular. A piece of 22-gauge (0·5 mm) sheet copper is cut to the required shape and drilled or punched to receive copper tacks with their heads almost touching.

A patch of thick canvas is cut to similar shape, leaving about 1 in (25 mm) excess all around. Bed the canvas to the hull in thick paint or mastic and locate with two small tacks near the middle.

Apply more dope, then tack the copper patch on top using tacks ½ in (12 mm) or ¾ in (18 mm) long, according to plank thickness. Omit any tacks which come in line with a caulked seam and fill any empty holes with stopping. Trim off the surplus canvas with a sharp knife and beat the tingle all over (particularly the edges) with a rubber faced hammer. A coating of antifouling paint is only necessary for the sake of appearance – copper has good inherent antifouling properties.

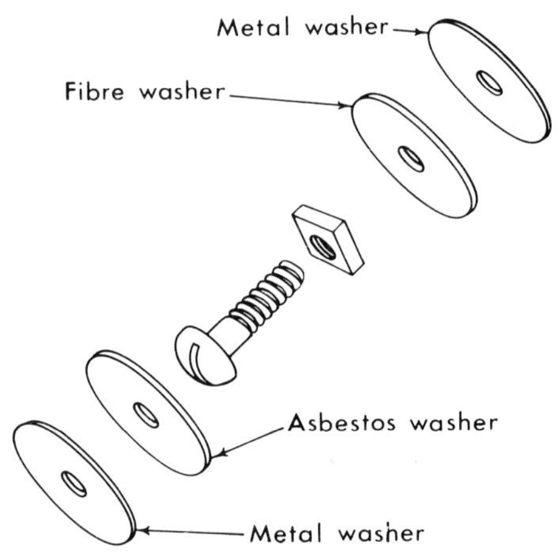

36. Saucepan-mender for sealing pierced steel hull.

Collision Mats

When a craft cannot be slipped, beached, or careened after collision damage, quite large tingles may be fitted by a diver. Even prompter action is possible if a collision mat is carried on board. You can buy specially made ones, but a 6 ft (2 m) square of heavy PVC-coated nylon is ideal if hemmed and eyeleted around the edges.

If serious leakage occurs after hitting a submerged object, two nylon ropes (heavier than water) are looped over the stem and see-sawed under the keel, enabling the collision mat to be pulled over the damaged area, PVC side to the hull. Other short ropes attached to the eyelets are made fast to prevent the mat from slipping due to the boat's movement. With any luck you can then stop pumping!

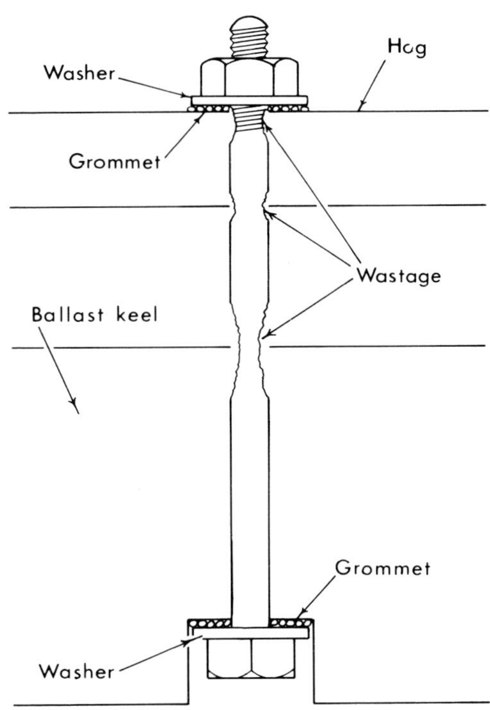

37. Keel bolt details.

Sealing Metal

Small holes through the bottom plating of metal craft, whether caused by corrosion or piercing, are easy to repair temporarily with fibreglass or saucepan menders. If your modern hardware store has never heard of the latter, they are simple to make (see fig. 36) preferably using offcuts of stainless steel for the washers and a 3/16 in (4 mm) bolt of the same material.

A tingle is readily fitted to a metal hull (see Chapter 9) by using pop rivets (or small bolts) on thin plating, and tapped holes with short set screws for plating thicker than about 3/16 in (4 mm). The patch should be of 16-gauge sheet metal compatible with the hull material.

Keel Bolts

Ballast of iron filings or lead shot encased within a fibreglass keel could escape if the keel gets damaged, so it pays to find out just what you are carrying at the bottom of the boat.

Similarly, craft have been known to lose their conventional external cast ballast keels at sea because of undetected keel bolt corrosion. When strange leakage from the keel region occurs in a boat more than ten years old, and the keel bolts are of galvanized steel, do not hesitate to get them X-rayed or withdrawn for inspection.

This problem is not limited to sailing craft, because many power cruisers have quite massive ballast keels.

Keel bolt wastage and leakage most often occurs where indicated in fig. 37. Sound bolts can leak if the grommet of caulking cotton under the head washer or the nut washer has perished. Getting the nut off to renew the top grommet isn't easy. If rusted, first clean up nut and protruding thread with a power wire brush and an old hacksaw blade. Dope it with penetrating oil four or five times throughout a full day.

If the nut is perfect (or can be shaped up with a file) borrow a heavy duty socket spanner which fits tightly. If the nut is misshapen, try a really big stillson wrench. In either case, extension piping at least 3 ft (1 m) long will be necessary to gain enough leverage. If no joy, heat the nut with a blow torch (after covering all adjacent woodwork with wet rags) and try again. The last resort is to break the nut off by drilling a few small holes up the middle of one face then splitting along this line with a cold chisel and heavy hammer. Borrow a die nut to refurbish the thread and fit a new galvanized nut.

A 1 in (25 mm) diameter bolt would normally need two strands of twisted cotton wound tightly around the bolt three times and then loosely spiralling outwards three turns. Bed this in mastic, rather than the traditional white lead paste mixed with tallow or motor grease. If the bolt hole was bored oversize, force mastic into the void and form an extra big grommet.

Floor Bolts

A surprising number of boats have galvanized floors fixed through the planking with copper clenches. Galvanic action sets up (see Chapter 4) and slowly softens the timber around each bolt, then leaks commence.

All is not lost when this happens, as quite successful repairs are possible by excavating all the affected timber and replacing it with epoxy putty. At the same time, replacing the bolts by stainless steel ones is often advisable.

Vibration

Engine bearer bolts rarely corrode, but they can work slack due to vibration, sometimes enlarging their holes through the boat's skin. Tightening the nuts may suffice, perhaps coupled with permanent tingles covering each head. When very slack through the hull, ease the bolts out sufficiently to apply the epoxy putty treatment.

Vibration causes trouble in other directions. Mounting bolts for engine, gearbox, thrust bearing, or sterntube may slacken, also the bracket fastenings holding half-floors and intercostals to engine bearers. These can lead to direct or indirect leakage.

Sometimes the seam stopping along planks close to the bearers shakes loose, so regular examination in this vicinity pays dividends.

Stern Glands

A few drips from the stern gland when an engine is running matters little, but leakage when stopped could prove serious. Beware of over-tightening a gland. This can cause overheating, though the same symptoms are sometimes created by misalignment or a bent propeller shaft.

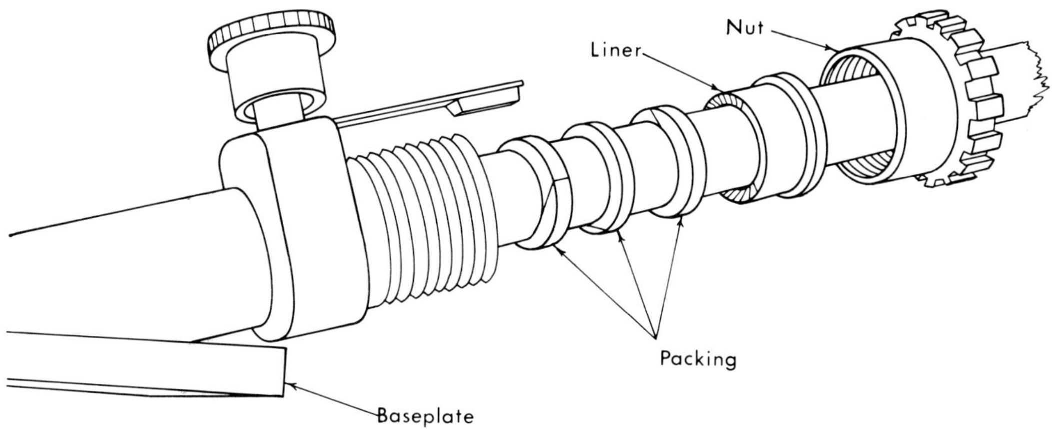

38. Parts of a shaft log type sterngland.

Gland re-packing is necessary when a gland still leaks after tightening or when it continues to run hot after slackening. Standard glands (fig. 38) are packed with rings of special square section rope pressed down by a removable liner. If the boatyard will not lend you their tool for raking out old packing, you may be able to scrounge a suitable discarded probe from your dentist!

Take a piece of old packing to the boatyard, chandlery store, or a plumber, to make sure you get the right gauge of new material. Wrap this around the propeller shaft and use a sharp knife to cut across both parts diagonally to form perfect rings for insertion into the stuffing box. Stagger the joints and fit sufficient rings so that the liner disappears about one-third of its length when the nuts are screwed down hand tight.

Some powerboats use patent oil seals instead of glands. If these leak, call for professional assistance.

Cathodic Protection

The ways by which dissimilar metals submerged in sea water can destroy each other by galvanic action was described in Chapter 4. Quite a few craft have copper fastenings and a bronze propeller, with iron ballast and a galvanized rudder. Others have bronze propellers with zinc sprayed steel or light alloy hulls, so the necessity for vigilance is obvious.

Livelier action concentrates where a small area of paint is scratched off than might arise with 100% bare metal. Further trouble is likely when a metal hull is moored close to a boat having copper sheathing or copper antifouling paint.

A simple precaution is to bolt sacrificial zinc anodes (plate 14) near to likely trouble spots. These are composed of highly anodic pure metal which is eaten away in preference to any other adjacent metal. The sacrificial part is generally screwed to a base plate which is bolted through the skin (or welded to a steel hull), with internal cables leading from one of the bolts to the sterntube, rudder, keel bolts, or the engine gearbox. Anodes are also made to clamp around propeller shafts. Others are attached to a length of wire for suspension from the deck while at moorings.

The matter is far from simple. If in doubt, consult a corrosion expert or naval architect. He may specify the more elaborate impressed current system for a metal or ferrocement hull. Any stray current leaking from a boat's electrical wiring can cause disastrous galvanic corrosion, so a battery isolation switch is always a wise precaution.

Nail Sickness

When an old boat leaks past many plank fastenings she may be *nail sick*. This malaise is caused either by corrosion (see earlier under floor bolts), vibration or flexibility when under way. In most cases all you need to do is tighten the slack fastenings. When this fails, new ones alongside may be necessary. Gaping parts rarely pull right home owing to the presence of dirt, but little can be done about that.

Riveted copper nails are easy to tighten by hammering inside with a dolly pressed on the head. This process is especially beneficial for tightening leaky clinker boats. Old woodscrews tend to shear off when tightened, so it pays to remove them instead and drive new ones of bigger gauge. Corroded bolts often shear off or rotate bodily before they will tighten – best to cut off the nut, drive out and fit new.

Drop Keel Trunk

Although a properly built centreboard or swing keel case should never leak, it receives a lot of punishment – particularly when running aground on a windward leg. With old age, a seam may start and curing such leaks is not always easy.

Sailing dinghies suffer the most, often because one fails to notice that the important joint with the centre thwart has failed. Without this connection, the hog fastenings cannot cope. There is often no way to get at these screws without removing some planking in way of the keel, so better not to try removing the case completely. Quadrant beadings (fig. 39*a*) with mitred ends, set in mastic and

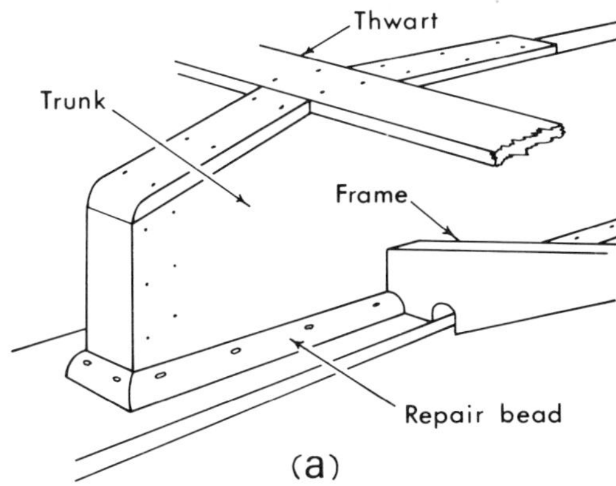

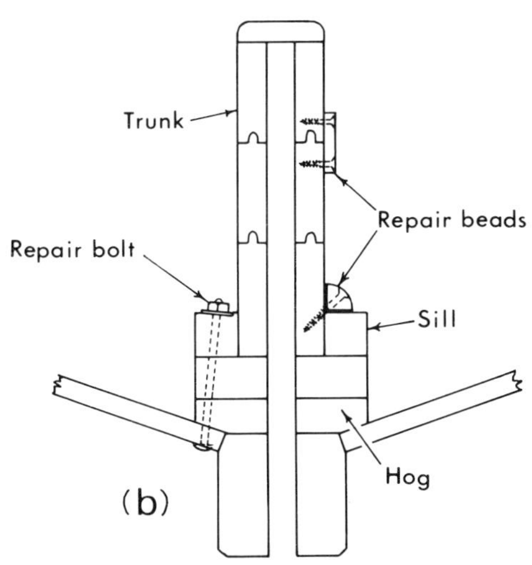

39. Drop keel case repairs.

screwed to both case and hog, will normally cure all leakage from this joint.

On bigger craft, you may find massive sills (fig. 39*b*) connecting trunk to hog. If the bolts refuse to tighten down, fit additional ones right through to the planking. Having secured the joint against further movement, the beading treatment should seal any leaks still persisting.

The planked sides of a big trunk should have splined seams which leak only if the timber rots. Caulking the seams from outside is possible in some cases, but the most effective way to seal leakage is to screw battens in mastic over the offending seams.

A galvanized steel case is more likely to buckle than to leak. Fortunately, complete removal is usually not too difficult. If the interior is badly rusted it might pay to have a new one fabricated.

Leakage from a drop keel pivot bolt is usually a simple matter of renewing the rubber gaskets. However, such bolts can wear alarmingly and early renewal is never a bad thing. A heavy plate must be jacked up to do this and a little juggling may be necessary to get the shoulder of the worn bolt free from the plate.

Portlights

The usual reason for leakage from opening portlights is simple failure of the rubber sealing ring. Some yachtsmen tolerate such leaks for years, stowing mattresses away from the drips, or hanging buckets underneath.

Rubber renewal takes but a few minutes. If the fitting is stamped with the maker's name, order new strips from them. Otherwise, order through a chandlery store, first checking the width of the groove and measuring its circumference by passing a piece of string all around the portlight bezel, then stretching the cord along a rule. Most strips are square in section, but try to check this by carefully removing a piece of the old rubber.

To insert a new strip, rake out all old material, trim the new piece to length making a tight butt joint, apply a few dabs of impact adhesive, press into place with the joint at the top, close the portlight and clamp lightly. Should a portlight or decklight leak underneath its mounting flange, complete removal is essential. Clean off all old putty, then replace in bedding compound. Fastenings are usually bronze bolts or woodscrews and should remove easily.

If there is a spigot through coaming or planking, hard mallet blows may be necessary to shift it, so it pays to drive out the hinge pin and remove the glazed bezel first. Floppy portlights are a menace – get the boatyard to renew a worn pin if you are not conversant with this type of work.

Even when free from leakage, portlights can drip from condensation. Shell-type *drawer pulls* turned upside-down and screwed (with mastic) directly underneath usually check this annoyance. A little cotton wool inside each one aids evaporation.

Windows

Leaky sliding metal framed windows are best taken off during the winter and returned to the makers for servicing; the same applies to fixed frame windows with rubber glazing. Leaks past frame flanges are generally easy to cure by re-bedding.

Windows glazed into a coaming rebate and secured with a metal bezel screwed to the outside often need a special gap-filling putty to ensure success when re-glazing. The two-part synthetic rubber compositions used for caulking decks (e.g. Life-Calk) work well enough when the special thicker glazing compound is not available.

Acrylic window panes (Plexiglass or Perspex) often need renewal due to crazing or surface scratching. Many are fitted into fibreglass or plywood coamings with special rubber beadings needing no frame, like a car windshield.

Engaging a firm specializing in the repair of excavator and tractor windows is often the best bet. For a popular mass-produced boat you should be able to obtain new panes with the correct length rubber beadings. When the latter are of the type that has a separate pop-in insert strip, you can do it all yourself quite easily. Those without an insert need expert handling to fold the outer lip over the coaming by means of a piece of strong cord wrapped around the groove. Smear ample mastic into the groove before fitting any type of glazing bead.

Some acrylic panes are simply screwed or bolted on with no frame or bead. When making new panes, always use a hand drill to bore the holes – the heating and snatching of a power drill can shatter this material.

Skin Fittings

There is little need to stress the danger of leaky skin fittings, whether for cooling water, heads, sink waste, bilge pump, cockpit drains, or other duties. Leaks around a bolted flange could be the result of broken bolts and that is serious. Re-bedding is similar to the portlight procedure. Even when apparently sound, a different bolt should be withdrawn every five years to check the general condition. Skin fittings are mentioned again in Chapter 10.

Leaks from plumbing, tanks, pump glands, or exhaust systems should not be neglected – remember the stitch in time proverb!

nine

External Repairs

When it comes to collision damage, metal hulls generally prove more rugged than those of any other material. Impact is absorbed by denting long before fracturing occurs. When jacking back or panel-beating is unsuccessful, parts of any shape or size can be replaced by welding or riveting – except perhaps in some remote parts of the globe.

Good quality fibreglass or ferrocement does not rot or corrode and can have better collision resistance than timber. However, the former does not stand up at all well to chafe and few boatyards know what to do with a ferro hull which has been battered sufficiently to leak.

Common Faults

Cabin boats are subject to much the same types of external damage as described for dinghies in Chapter 6, but they suffer far more from rot and corrosion, galvanic action (see Chapter 8), chafe against wharf piles, damage from anchor flukes, running aground, falling over, and burning.

The repair of toe-rails, sheer beadings, stem, rubbing strakes and coamings is similar to dinghy work, so we will now concentrate on the outer skins of hull and decks in all the materials commonly used. This chapter should be studied in conjunction with Chapter 6.

Most wooden boats are now of a considerable age and as timber is bound to deteriorate in time, repairs occasionally become necessary irrespective of accidental damage. Fortunately, timber repairs are usually simple and well within the capabilities of most amateurs; they are not too expensive and are enjoyable to do.

Carvel Planking

Any plank, or part of a plank, is replaceable in a traditional boat (plate 18) without need to disturb the adjacent strakes. Chop out the faulty piece, fashion a template from hardboard or scrap timber, shape the new plank from that and fasten it in place.

If the old piece is removed intact, this will act as a template, eliminating one stage. Sometimes the new board can be held in position outside the hull and the shape scribed direct on to it from inside. However, due allowance is necessary for the caulking Vee and the board may not bend properly around a bulbous hull without steaming, even using jacks from a wall or vertical posts pulled inwards with Spanish windlasses. Long pieces obviously bend more readily than short ones, but planks nearly always have a certain amount of sweep (see fig. 40*a*) and the narrow final plank will bend more readily than the full width board one cuts it from.

Cramping short planks when out of doors is particularly difficult. One good way is to screw bridging brackets across the adjacent planks (fig. 40*b*) and drive wedges under

Plate 18. A new piece of carvel planking being caulked with oakum.

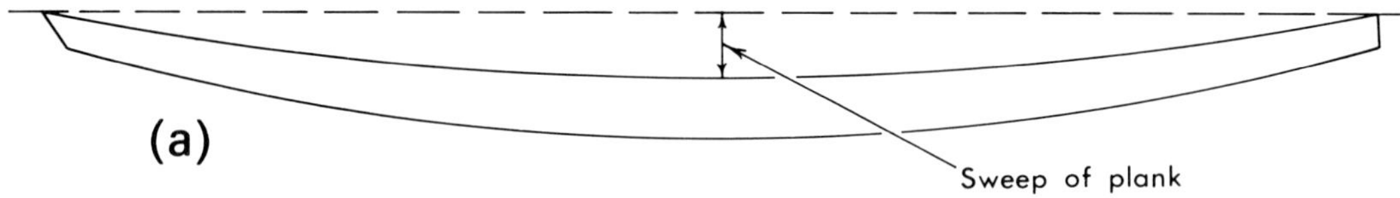

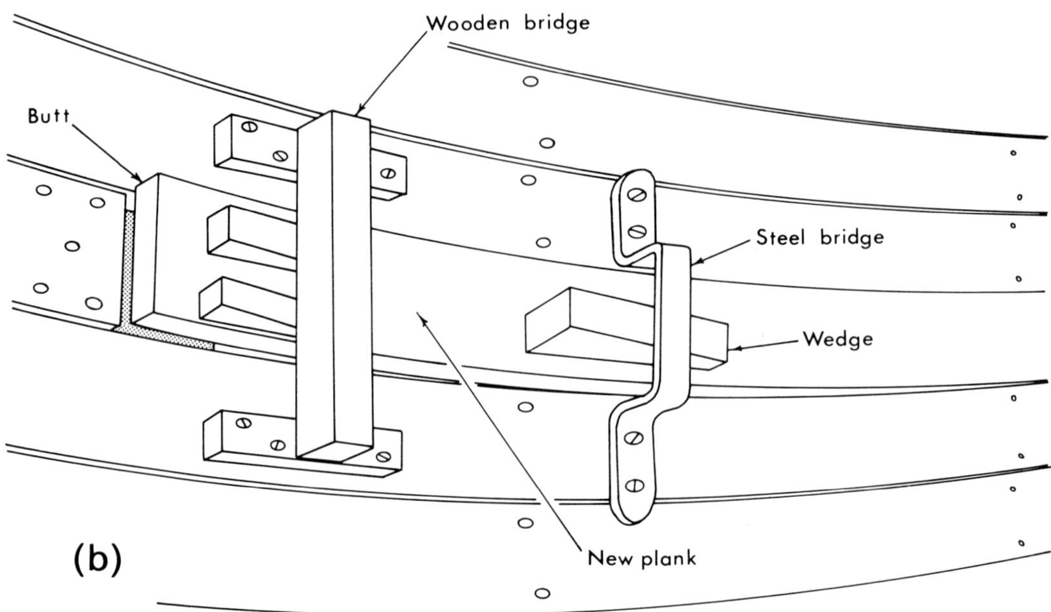

40. Carvel planked hull repair.

these.

Do not let the butt ends of a repair piece join on a frame except for heavy double sawn frame construction. Instead, fit ample butt blocks inside overlapping the adjacent planks a little and well fastened to old and new timber.

For a painted hull, a short repair to a single plank (perhaps spanning just beyond two frames) is quite satisfactory. To obviate steaming over a jig, the repair piece may be cut from thicker wood, then planed hollow on the inside and convex on the outside to form the necessary curvature.

Short repairs look terrible on a varnished hull. Renew as long a length as possible to camouflage the butts, or change the hull to a painted finish.

Clinker Planking

Repairing a clinker (lapstrake) plank as in plate 19 is a little more tricky than for carvel work, as the new piece has to tuck under the plank above it and lap over the one below. Particularly at the turn of the bilge there will be some bevelling to do (fig. 41*a*) plus some rabbeting at stem and transom – see fig. 41*b*.

You cannot fit butt blocks efficiently in clinker work due to the lands (overlaps), so all butt joints should be scarfed. Incidentally, scarfing is often better for carvel repairs if butt blocks would look unsightly for ever more inside the boat.

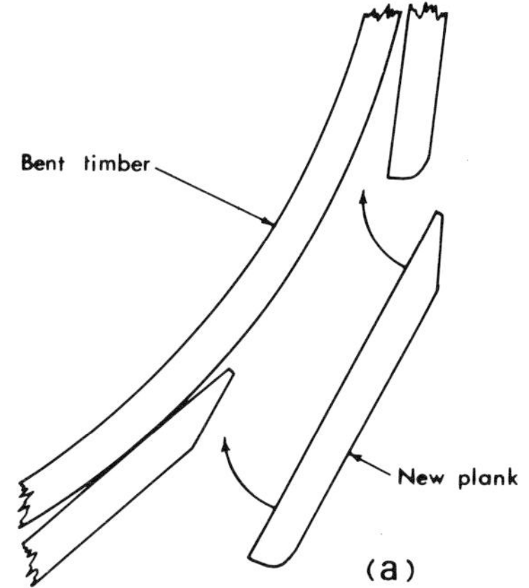

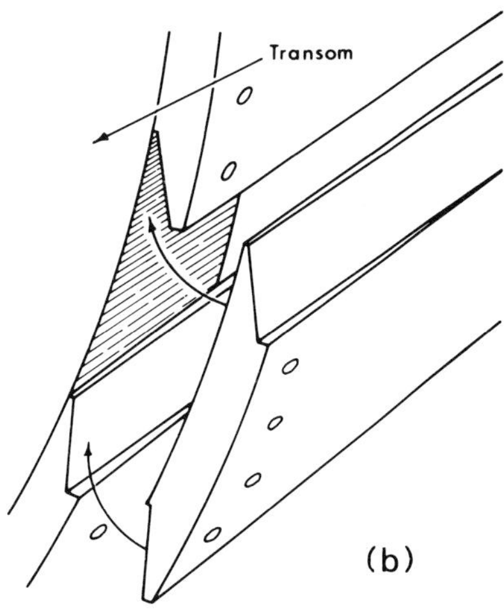

41. Single clinker plank replacement.

Plate 19. Any part of a clinker plank can be renewed without removing adjacent strakes.

A horizontal template will pick up the scarf angles if necessary, but normally you only need to transfer the old plank shape to the new timber then measure back from each end to mark out the scarfs. Provided the new wood is the same thickness as the original, scarf angles are bound to come out right. Though it may not be accurate enough for gluing, a dollop of mastic is good enough if you set two rows of copper rivets through each scarf.

One benefit with clinker – the copper fastenings are usually easy to remove, see Chapter 6. If extraction is carefully done, the nail holes may be used again, perhaps with nails a size larger to ensure tightness.

Double Diagonal

Extensively used on ex-naval craft (many of which have been converted to yachts) and powerboats of all types prior to the advent of plastics, double diagonal planking consists of two thin skins of planking crossing at right angles, each laid at about 45° to the waterline, with a layer of oiled calico between them. Numerous copper fastenings all over ensure a strong and tight hull.

When the inner skin rots, the outer planks must be removed to get at it, so repair work is by no means simple. New butt joints should be well staggered (fig. 42), then little strength is lost by using one skin as the butt blocks for the other.

The inner skin goes back first. You will probably need to screw short temporary stringers between frames to ensure the skin assumes true curvature. Screw small temporary butt blocks inside to stop the plank ends from springing outwards till the other skin is laid.

Every other strake usually has parallel edges, while the intermediate ones are slightly barrel-shaped. Mark these by strutting the plank in place while overlapping its neighbours, then scribe the edges from inside. Naturally, you cannot do this with the outer planks, but for a small repair job it does not take many minutes to plane up hardboard templates by trial and error.

Calico (or a piece of 4 oz cotton sailcloth) will do in between, but lay this in thin mastic over and under, overlapping the old calico by about 2 in (50 mm). Bed all seams in mastic.

Using the same type of fastenings as before, a repair job should be invisible inside after painting.

Moulded Wood

The modern cold moulded system is similar to the above, but all surfaces are glued and there may be as many as five layers of planking at various angles to each other on a large vessel. To repair a big hole, work from the outside, cutting back each successive layer in similar fashion to fig. 42.

Fit shaped backing boards over the inside of the hull, covered with cellophane and strutted into position. Layer by layer, fit new pieces of wood (the same thickness as the originals) into the gaps, gluing mating surfaces and edges. Use staples or small pins to bed them down. After hardening, remove the backing and grind off any protruding pins.

It often pays to shape all these pieces, offer them up and number them, then complete the gluing in one session. Otherwise, the glue oozing past each layer must be carefully wiped away before hardening. If short pieces prove too stiff to take the curvature, use two veneers of half the original thickness for each layer.

Strip Planking

This system comprises a single layer of fore-

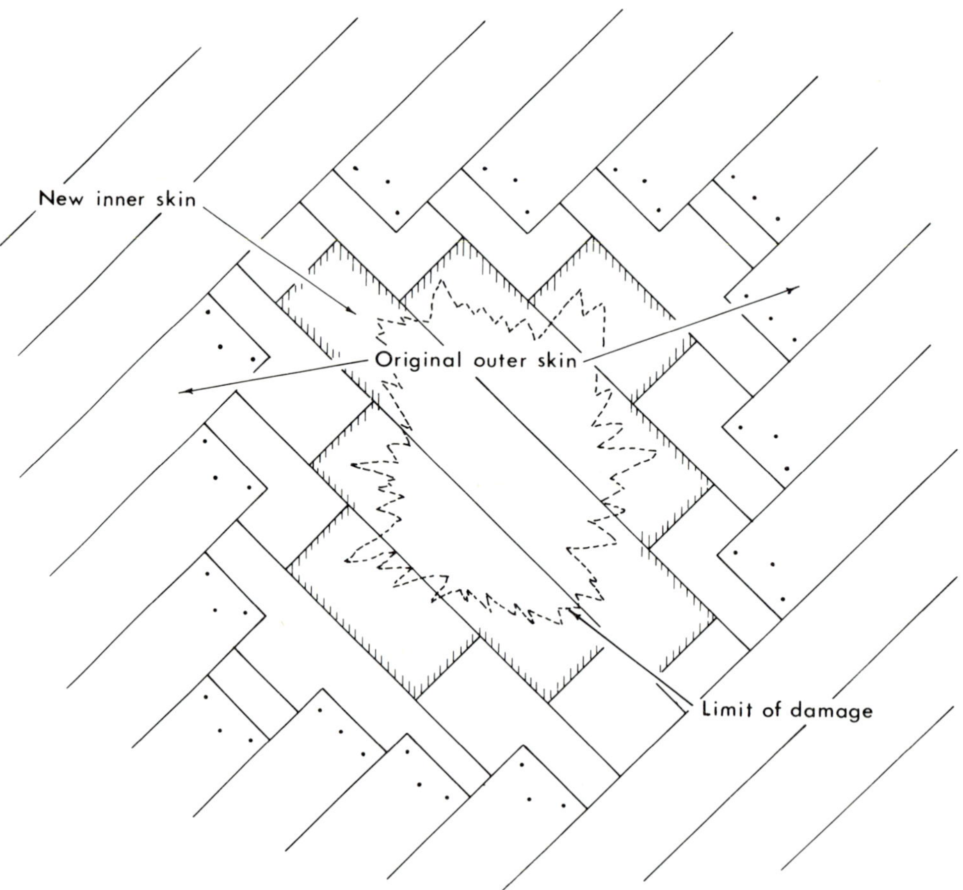

42. Double diagonal repair – original fastenings omitted for clarity.

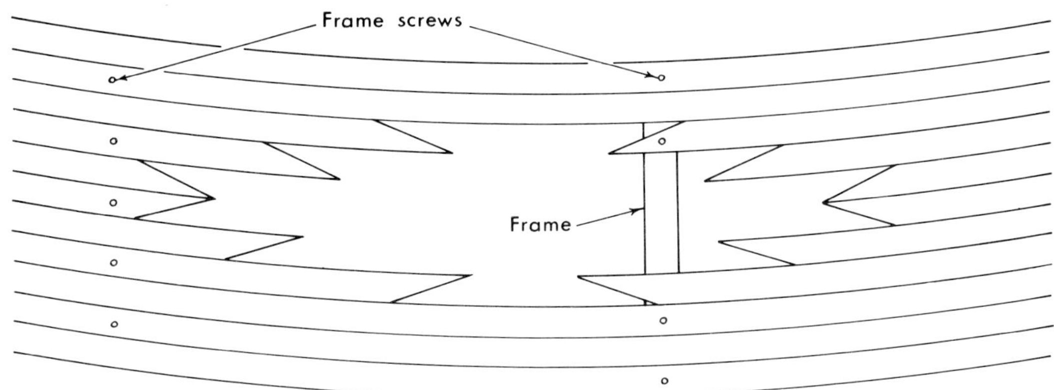

43. Scarf layout for strip planking repair.

-and-aft battens little wider than their thickness, sometimes glued, sometimes set in thick paint, but always pinned together with vertical nails passing through at least two strips.

Cut away the damage as shown in fig. 43, then start at the bottom, shaping each strip by planing to the inevitable curvature both ways. Cold bending may be possible for a long repair.

Bore pilot holes for the new edge nails whenever possible and use a big piece of heavy flat bar and a hammer to drive the nails home when space gets restricted. Pin the final strips by boring from outside for short oblique nails upwards and downwards, punching them well in and making good with stopper.

Repairing Plywood

Wherever possible, repair big holes in plywood topsides, bottom, or deck, by cutting away the damaged area into a neat circular aperture, or a rectangle with well rounded corners. Shape a new piece to fit into the aperture, then glue a pad of similar plywood with generous overlap on the inside. Glue the outer panel into place and make good around its edges with stopping. If the fit is very sloppy, use epoxy putty.

Any number of permanent or temporary fastenings may be used. When there is much hull curvature, laminate each part with two or more skins of thinner plywood. Alternatively, steam or boil each panel and bend it over a jig before fitting.

Invisible repairs on varnished plywood are not possible without renewing a whole panel – see Chapter 6. When a flush interior finish is required, the procedure is similar to cold moulded repairs. Form a series of stepped rebates – three will normally suffice – then shape the repair panel to fit this, or laminate in situ with veneers of the correct thickness.

For very thin plywood, rebating is difficult, even with a router. Ordinary tapered scarfs cut with plane and chisel are then best, using a temporary backing piece to preserve the feather edge. When rebating thick plywood, gauge the rebates to match standard thicknesses of thinner plywood.

Graving

Small defects in timber caused by knots, abrasion, or rot are normally made good with epoxy putty – or by tingles below the waterline, see Chapter 8.

As putty ruins the appearance of varnished work, the traditional method of repair is by graving – inlaying a patch of matching wood, its thickness about one-quarter that of the planking.

A graving piece should be diamond-shaped, but longer with the plank grain than across it. Glue only should suffice to hold it, though a few panel pins are permissible, punched well down and stopped over. Make the piece a bit thicker than the chiselled recess and bevel its edges slightly to form a good tap-in fit. Plane down flush, stain if required, then build up the varnish coating.

When a plank is pierced right through, fill the hole beneath the graving piece with epoxy putty.

Plastics

Nowadays the shipwright's art is almost forgotten, but nearly everybody knows how to work with fibreglass. All sorts of stores sell repair kits, many containing all the instructions necessary for use.

Some fibreglass hulls left afloat continuously have developed hundreds of blisters over the underwater gel coat – the problem of osmosis. The standard remedy for this is to grind off the entire gel coat, allow several months for the matrix to dry out, then apply seven coats of two-can polyurethane paint.

A fibreglass boat with extensive hull damage is unlikely ever to get repaired. Unless the original mould is still in existence and available for making a new hull section to glass into the otherwise sound hull, no boatyard is likely to take the job on at an economic price. However, a keen amateur might just be able to borrow an identical boat for long enough to take a mould off her. Alternatively, if damage is restricted to one side of the hull, one can build a wooden male plug (replica, or mock-up) of the damaged area using measurements taken from the opposite side, then make a female mould from that. A clever amateur could make a good enough female mould in wood to eliminate one stage.

Repairing small defects presents no problems, but anything larger than about 12 in (300 mm) square needs much care to ensure a true blend of hull curvature. This is generally achieved by taping and strutting a sheet of thick cardboard (covered over with cellophane) to the outside of the hull, gel coating this and then applying the necessary layers of impregnated glass matt or cloth from inboard.

Quite a number of fibreglass patches have fallen out after a year or so. A good key between old and new is therefore essential. Bevel the raw edges with a rasp as in fig. 44*a* and use a coarse sanding disc ruthlessly on all other bonding surfaces.

A bulge inside (fig. 44*a*) adds greatly to the strength of a patch and stops it falling out. One can attempt a flush repair (fig. 44*b*) which is invisible inside the boat, but its strength – except perhaps with really thick laminates – may always be a little suspect.

True colour matching with a pigmented gel coat is not reliable in white and almost impossible with other colours. If a topside patch looks offensive, paint the hull at least on that one side.

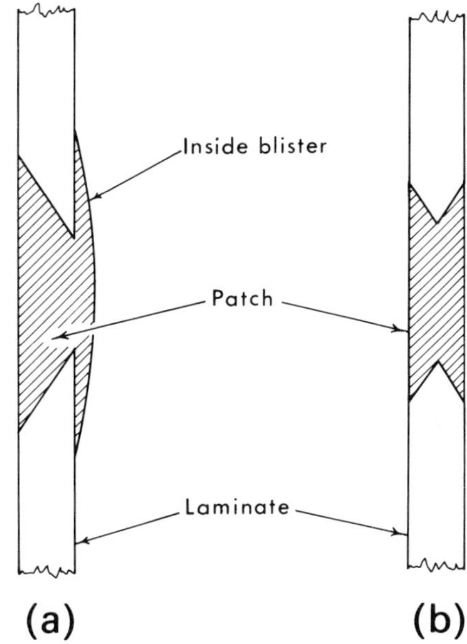

44. Fibreglass hull patches.

Plastics on Wood and Metal

Patches of fibreglass work quite successfully on wooden hulls, particularly cold moulded and strip planked ones. However, with traditional planking, the seasonal timber movements could upset a patch covering more than one plank width. A double rebate (similar to fig. 45*a*) stops a patch falling out, but a fibreglass graving piece held in with wood screws is often just as good – see fig. 45*b*.

Epoxy resin adheres better than the polyester of common repair kits when using fibreglass to patch metal hulls. Even then the metal must be thoroughly roughened and

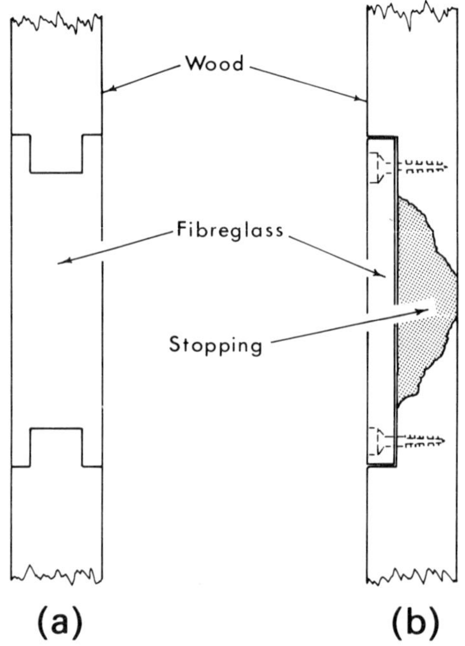

45. Using plastics to repair timber planking.

work carried out in ideal weather conditions.

Good repair kits contain adequate literature on how to use the materials, also a table showing the size of kit required to build up various thicknesses over a certain area. The correct amounts of chopped strand mat, surfacing tissue and woven glass cloth should be included to absorb the quantity of resin supplied on an average repair job.

Ferro Skins

It takes a big bash against a sharp rock to bruise a ferrocement hull badly enough to need remedial measures more complicated than epoxy resin. Any exposure of the metal mesh armature must be sealed immediately, even though there may be no leak. Special epoxy resins are available for concrete repair work – one for brushing on the old surface before building up with mortar and another to fill small defects and cracks.

Some ferro hulls show hair cracks all over due to hurried curing, mainly externally. A really sound covering of epoxy or two-part polyurethane paint should master this.

Any breakage right through the hull or deck needs careful treatment to avoid future weakness. Chisel away all loose mortar both sides. Holding a heavy dolly inside, hammer the mesh from outside and blow frequently with an airline nozzle at 100 psi to clear as much mortar dust as possible. Alternate the process frequently, holding the dolly outside.

If the damage is bigger than about 9 in (230 mm) across, cut through each layer of mesh in varying directions from the middle and peel it back to permit further hammering. Take care not to disturb the main bars unless already broken during the accident. Any broken bars should be sistered with a piece of the same steel, well overlapped vertically and laced with 20-gauge plain wire.

Dress back each layer of mesh with a rubber-faced mallet and bind all layers together with lacings of wire threaded through, pulled tight and twisted.

Brush the special epoxy liberally on to the old exposed mortar faces. Hire the services of a good plasterer to trowel off outside while fresh mortar is forced right through the armature from inside.

When the surface is fair and smooth both sides, seal off completely with a double layer of polythene sheeting taped inside and outside the hull in airtight fashion. Leave undisturbed for two weeks and keep the hull out of the sun if possible.

The mortar mix is exacting. Best to obtain a pre-mixed bag from a ferrocement specialist. If this is not possible use only sharp washed silica graded sand, all passing a $\frac{3}{32}$ in (2·36 mm) sieve. Mix $1\frac{1}{2}$ parts sand with 1 part of fresh portland cement, adding only sufficient water to create a workable consistency.

Patching Steel

Big dents in steel skins may need to be pressed out and hammered to achieve a surface good enough to finish with a thin skim of epoxy putty.

Thin plating may be panel-beaten as for a buckled car. Around $\frac{3}{16}$ in (4 mm) plate needs a sledge hammer one side and a heavy baulk of timber the other side. When using a hydraulic jack to press out a dent, take care that the load is well distributed inside the boat. It could prove safer to make a small hole

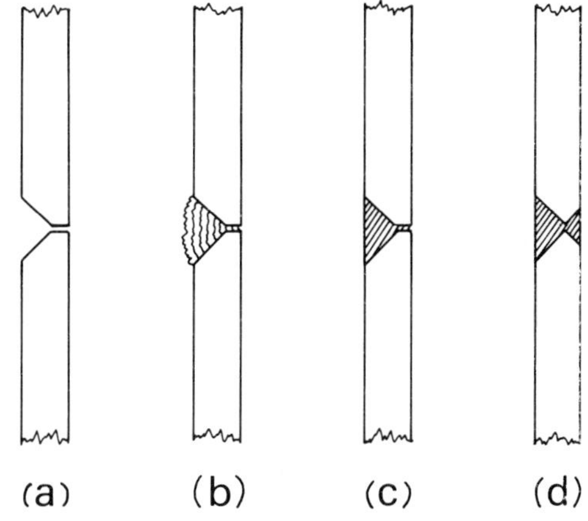

46. Butt weld for thick plate.

through the plating, put a bar inside and apply a pulling winch from the outside.

With thick plating, or where a frame is buckled also, it generally pays to cut away the whole of the damaged area and weld in new steel. Fit the skin plating first, then make a template for the frame.

Welding

To cut out a section of skin, use a power *nibbler* for thin plate and a gas cutting torch for thick. Make the patch a slack fit and align in position for tack welding by means of magnets.

Alternatively, bore holes each side of the seam and bolt across bridging straps of steel or wood.

Little edge preparation is needed for thin plate. A small gap is left between the old and new metal and later built back with weld. Upwards from $\frac{1}{4}$ in (6 mm) thickness, full bevels (fig. 46*a*) are necessary. Sufficient gas or electric arc weld runs must be made to fill the Vee (fig. 46*b*) and to permit grinding flush (fig. 46*c*) if necessary. A small run of seal weld inside (fig. 46*d*) improves the appearance, especially if also ground flush.

A certain amount of distortion is inevitable when welding and is generally accepted in the finish of steel workboats. Most steel yachts with mirror-finish topsides carry around an awful lot of epoxy putty!

Riveting Steel

When unfamiliar with riveting, practice on test pieces until you know just how much to clip off a rivet tail and how to obtain a neat closure. Steel rivets up to $\frac{3}{8}$ in (9 mm) diameter can be worked cold, permitting repair patches to be bedded in mastic. Bigger steel rivets need heating to redness; this has the added advantage of extra tightness by the contraction on cooling. If new and old plating is not zinc coated, paint all mating surfaces and allow to dry before riveting. Finally, all edges can be caulked by peening. If you cannot borrow the correct tool for this, grind an old cold chisel with a slightly bevelled flat tip and drive this systematically along the plate edge using a heavy hammer.

Accurate rivet boring is essential, so pre-drill the patch and fix it into position with a few bolts while using it as a template to bore all remaining rivet holes.

A flat dolly is fine over the heads of countersunk rivets but for round heads (where appearance matters) borrow a set or *snap dolly* having the correct size hollow in its tip to fit the rivet snugly.

Plate Dishing

To patch a round bilge hull having compound curves, the new piece may require *dishing.* Make a few templates from thin scrap timber for various positions across the repair. If the curvature is mainly in one direction, common bending rolls may produce a fair shape. To get a compound dishing without shipyard equipment, hammering all over along chalked grid lines will produce the shape needed in plate under $\frac{1}{4}$ in (6 mm) thickness.

Lay the plate on a block of hardwood. Use a heavy ball pane hammer for thin plate and a blacksmith's *fuller* struck by a sledge hammer for thick plate. Don't forget to wear earmuffs!

If you over-do it, beat the convex side with a flat hammer while holding the piece on top of a timber post. Start with an oversized plate and trim the edges only when the templates fit correctly.

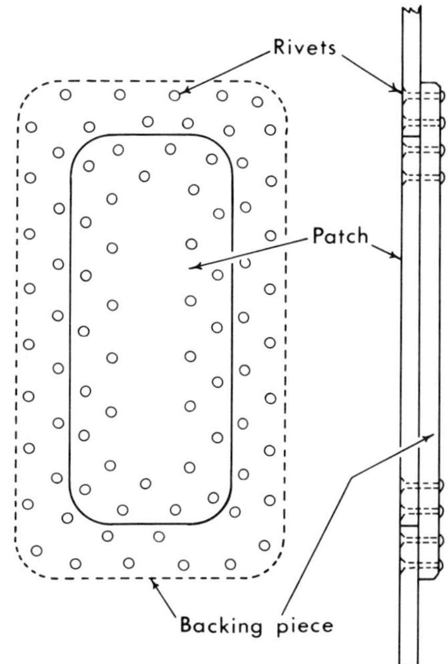

47. Flush riveted patch in alloy skin plate.

Light Alloy

Pushing out dents in marine alloy hulls is simpler than with steel, but the metal may have stretched so that it will not go back to the original position without buckling. This can be partially masked on a polished hull by painting it and using trowelling cement where necessary. Otherwise, it means annealing by careful heating followed by expert panel beating, then buffing back to bright metal.

The amateur cannot be expected to weld marine alloys reliably, but riveting is far simpler than with steel and should suffice for most needs.

Two rows of staggered rivets are always better than one straight row, and the use of butt plates inside (fig. 47) ensures that a patch is flush to view. Aluminium rivets are always worked cold and should be closed with a minimum number of hammer blows, as the metal hardens with repeated hammering. Pop rivets inserted with a plier-type or scissor-type tool are widely used with alloys, but they cannot normally be buried with a flush finish.

Only by stopping and painting can invisible repairs result in aluminium, but many people prefer this sort of finish in any case. It also enables fibreglass repairs to be used in future for making good small dents and holes.

Centreplates and Rudders

A buckled cruiser drop plate is fortunately a rare occurrence. It may refuse to hoist back even with the help of a diver and must be dropped in shallow or tidal water for recovery and repair. Straightening with a hydraulic press is easy, but a new plate is sure to be less liable to buckle again.

A bent metal rudder blade can often be straightened without removal by improvising a press made from baulks or steel channel each side connected by bolts, plus a jack and packing pieces. Dinghy-size plates are often straightened successfully by running a car wheel over them, followed by a little judicious hammering!

A cracked wooden dinghy rudder blade is often best repaired by enlarging the crack into an elongated Vee, then gluing in a spline of matching wood. For cruising boats, a reliable rudder is such an important safety feature that additional reinforcement is prudent if the above type of repair is attempted.

This is most easily done by inlaying flush metal straps in pairs port and starboard, held by woodscrews carefully positioned to avoid each other, or by copper rivets driven right through.

Another way, which is not quite so difficult as it might appear, is effected by boring edgewise into the rudder blade and inserting one or more bolts as in figs 32 and 48.

When the blade is thicker than about $1\frac{3}{4}$ in (44 mm), *drift bolts* are quite suitable; these are like enormous nails (fig. 48) having heads formed by riveting over heavy washers. A slightly different sort has a screw thread at the end which is protected by a cap when driving. With the cap removed and a washer and nut substituted, tightness is assured.

Drift bolts are driven into slightly undersized holes. If galvanized and more than $\frac{1}{2}$ in (12 mm) diameter, the zinc coating normally gives just about the right tightness when using a standard drill the same size as the original bar. A trial run in a piece of scrap timber is advisable, splitting the wood asunder afterwards to recover the drift bolt. To reduce the hole size, grind the outside of the drill a little. If the hole proves too tight, slight reaming is possible by quickly sliding in a piece of standard black bar with the tip heated to redness!

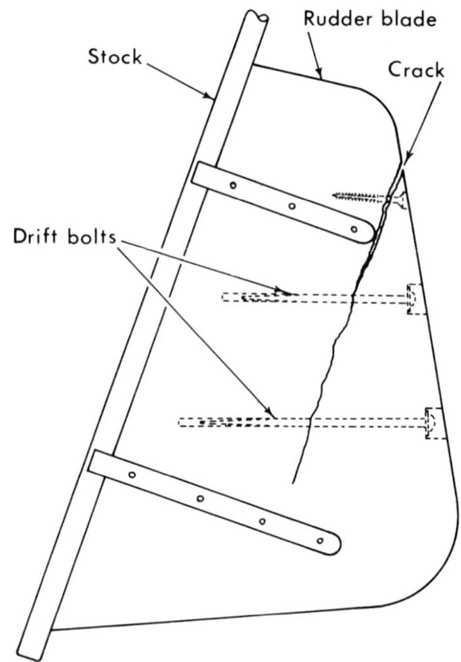

48. Drift bolts for a cracked rudder repair.

Rubbing Strakes and Bulwark Capping

Vessels of fishing boat type (and working craft) often have heavy duty timber fendering, sometimes capped by a half-round galvanized steel or brass strip. The more frail boats nearly always sport a sheer beading to help protect the topsides, while hulls having tumblehome aft often protect the bulge with a short rubbing strake in that vicinity only.

Overhanging strips are fitted to chines for protection and spray deflection. Deep-Vee powerboats adopt similar deflectors below the waterline, while shallow but long bilge keels on power cruisers help to check rolling and prevent a great angle of heel when the boat dries out.

All these protuberances need occasional repair. Few are structural parts and it just takes a little nerve to whip a piece off and apply the procedure described in Chapter 6.

Heavy timber fendering needs sawing to shape and there may be curvature both ways. Steaming is generally quicker and cheaper provided one can lay hands on the equipment. Regardless of the original method of fastening, through-bolts work best and help in adjusting small errors. Short lengths of oak or elm function quite well, but for a varnished finish a good match is necessary. When renewing a complete varnished strake, it pays to go for teak, iroko, or afrormosia.

Laminating saves wastage of costly timber and simplifies bending. Bolt heads should be countersunk deeply and stopped or dowelled even when covered by a metal band. Remember to paint or varnish all hidden parts.

Many steel boats have timber fendering. The bigger ones are secured by vertical bolts passing through continuous horizontal strips welded to the hull top and bottom. This system simplifies repairs as the bolts are easy to drive out and the new timber does not need to be a perfect fit against the hull.

Yachts with low bulwarks around the decks (perhaps with guardrails on top) are by no means out of fashion. The capping rails, often of varnished teak, are the most vulnerable part. Generally cut from the solid, frequent scarf joints are common, so it pays to renew a whole section when bad damage occurs. If the scarfs were glued, remove any fastenings through them, saw straight across and remove the damaged section of rail, then saw along the original scarf lines.

Rarely can the broken piece be used as a pattern, so make a vertical as well as a horizontal template and get a boatyard to saw out the new piece for you. Make card templates of the end scarfs and strike in their exact positions – together with the positions of any stanchion tenons – by standing the new piece on top. Do not plane the rail to shape until successful scarfs have been cut.

Slack scarfs look terrible on deck, so aim at a fit which is about $\frac{1}{16}$ in (1·5 mm) tight. Once you have a good fit at each individual end, chisel away the spare bits so both slip in together. If there must be an error, make sure it happens at the bottom.

ten

Improvements

No book on boat care and repair would be complete without a few tips concerning the improvements and alterations so often contemplated by the handy owner who has no desire to change boats frequently.

Appearance

In general, the smaller the boat, the uglier she is likely to be. Small cruisers need raised cabin top and high topsides to gain headroom. As many fittings are identical for craft of all sizes, the small cruiser appears to have guardrails like jail fencing and portlights like owl's eyes. A dinghy stowed on deck seems to cover half the boat.

With a little effort one can reduce the effect of high coachroof coamings by changing the quadrant mouldings at top and bottom from varnish to deck paint. Adding a deep boot-top band plus a coloured cove line or sheer band will take away some of the high freeboard expanse.

Adding a short bowsprit (fig. 49) improves the performance of many a sloop rig and gives an impression of greater overall length. It can also incorporate the anchor cable fairlead with advantage and you might like to fit a fashion board beneath the bowsprit to imitate a cute fiddle bow.

Adding teak toe-rails around the deck of a yacht nearly always improves appearance, and is also a good safety measure against the loss overboard of tools and gear – as well as people. An old sailing boat might look better with new pulpit, pushpit and guardrails.

49. Added bowsprit helps transform a sloop into a cutter.

Some craft look bigger and more workmanlike with fixed stern davits – see plate 20. This could solve the dinghy stowage problem, for even when an inflatable boat is used few owners take the trouble to deflate this for short passages.

Portlights with wide brass or chrome rims

Plate 20. Stern davits solve many problems and can look good.

need less polishing and often look better if these metal parts are painted to match the coamings.

Many an old boat looks rough because repair works have been carried out shoddily. The handyman can often take advantage of this, buying cheaply, then starting improvements which increase the boat's value while also providing an enjoyable hobby.

Joinery

Some modern fibreglass cruising boats have a surprisingly poor standard of internal plywood joinery. Renewing this item by item with high class woodwork makes an interesting long-term project for the enthusiast. This creates a good subject for conversation when friends are on board and adds to everybody's pleasure when cruising.

The parts most frequently needing attack are berth fronts, lockers, shelves, deckhead covering, hull lining, cabin and cockpit soles and drainage gratings. Some re-designing of the layout could lead to greatly improved galley, toilet, hanging locker and chart table facilities. There is an endless list of useful additions applicable to most craft. Does the cockpit have a proper helmsmen's seat? Are there neat racks for binoculars and cups? Is there a proper locker for gas cylinders vented overboard?

Condensation is a curse on many craft of plastics, metal or ferrocement. With the outlay of much time (but little expense) the important parts can be treated by adding insulated linings, cork paint, or expanded foam, plus improved ventilation and space heating.

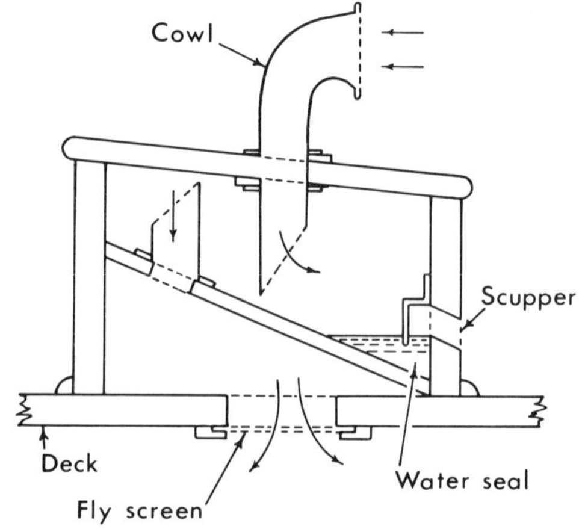

50. Section through a home-made dorade ventilator.

Ventilators

Few boats are adequately ventilated when unoccupied, yet too much draught is always better than too little – even if some rain gets in.

Good water-trap vents (some incorporating decklights) are readily available and are easy to install. Screw-down mushroom vents can be sealed for rough weather. They are virtually vandal-proof when open, but they are notorious for causing rope snarl-ups.

Even worse rope troubles occur with traditional stand-up cowl vents, and a useful improvement could be to replace old metal ones with the modern flexible rubber variety. An advantage of cowl vents is the ability to rotate them as desired according to wind direction. Mounting them on dorade water-trap boxes (fig. 50) makes a useful carpentry exercise. Concealing the boxes inside or outside coamings, against hatches or halyard bins, or just under deck at the stemhead, is an interesting project for ingenuity.

Improvements for hot weather conditions include making wind scoops for portholes and hatches, also awnings and canvas covers to protect brightwork.

Safety

Few boats are devoid of hazards which thoughtful improvements can eliminate. These range from slippery decks or dinghy bottom boards, inadequate guardrails, hand-grips, harness jackstays, or berth flaps, to fuel and gas leaks, sharp corners on cabin joinery, lack of padding on low deck beams, or broken strands of wire rope.

Do your navigation lights have sufficient power? What action will you take if the mast-head light bulb burns out? Is there a spare tiller for a tiller-steered boat or an emergency tiller in case your wheel steering breaks down? Do your portlights flop about when open, to cause eye and scalp injuries? Does the toilet door slip-bolt have sharp corners to gouge one's fingers?

Equipment having a direct bearing on safety can usually be added or improved, including items like a dinghy's compass, paddle, tow-rope, bailer, rowlock lanyards and buoyancy, or a cruising boat's fire extinguishers, flares, medicine chest, liferaft, spare bilge pump, or distress transmitter.

Lightning damage to yachts is, thankfully, rare. However, on all except metal hulls, fitting a copper strap inside from a masthead shroud chainplate bolt (or the base of a metal mast) to a keel bolt could save extensive damage.

Chafe

You need to be a jump ahead of chafe. It occurs mainly in the rig, where sails and sheets rub against stays, lifelines and spreaders; halyards always settle in the same position over sheaves which are too small; clawrings rub on reefed sails; unfrapped halyards beat against the mast incessantly.

The chafing of mooring lines through fair-leads and fender lanyards over toe-rails is easily prevented by wrapping with old canvas, or sleeving with pieces of hose pipe. Adding a bow fender will help to prevent anchor and mooring chains from damaging the topsides. A dinghy without proper gunwale fendering can soon scrape paint from the parent vessel.

Mechanical chafe is not uncommon. Rudder hinges (gudgeons and pintles) can wear alarmingly at moorings when someone forgets to lash the tiller; a propeller shaft can keep rotating slowly in a tideway; steering wires can rub against frames or faulty sheaves; fuel lines can vibrate, and wiring can fall across moving parts.

Inventory

Most sailing dinghy owners carry a mental list of all gear to be kept on board. The larger the boat, the more involved becomes the inventory list of instruments, galley utensils, stores, books, tools, spare parts, medicines and other equipment. Knowing where each item can be found is often a mystery, particularly for crew members or charterers. Not all yachtsmen (or even yachtswomen) are methodical in their leisure time!

If a written inventory is prepared one rainy day, divided into named lockers, drawers and other compartments – then kept up to date as necessary – this could save a lot of wasted time and might prove a useful safety measure in an emergency. By comparing your inventory list with those of other owners, some forgotten items are sure to come to light.

Should you sell the boat, a long inventory list is impressive. It should be easy to convince a prospective buyer that all those items would cost a great deal at present day prices.

Sheathing

Copper sheathing below the waterline is still used on wooden yachts cruising in tropical waters. It wards off marine growth and borers, the saving on antifouling paint generally paying for the copper in about eight years.

Nowadays, the amateur carrying out improvements is more likely to adopt fibreglass or nylon sheathing. This is cheaper, protects the planking, stops leakage, prevents a wooden hull from taking up water, has a good effect on galvanic action and stops marine borer activity. But it does not prevent weed or barnacle growth.

Avoid the temptation to sheath a leaky old tub. If she works under way, the sheathing is certain to break adrift from the planking in places and rot could start there. Excellent surface preparation is needed (plus complete dryness) for successful sheathing; this is difficult to achieve on a very old boat. In fact, modern sheathing methods work best on new hulls, when compatible stopping can be used for seams and fastenings. For an old hull, any putty should be raked out and replaced by the sheathing maker's recommended stopping. Any good chandlery store or boatyard will order the materials for you, together with instructions.

Hull Openings

If the improvement programme includes a self-draining cockpit, bottled gas locker, a sink, or certain electronic equipment, you may well need to fit skin fittings through the hull. Whether above or below water, this operation needs care – not only to avoid leaks but to guard against galvanic action. Fittings of inert plastics are now well-proven; bronze ones must not be used through a metal hull. Also, as the steel armature of a ferro-cement hull will be exposed, it pays to use plastics – or else surround a bronze part (and its bolts) with Delrin (nylon) bushes and washers.

Flexible piping or connections are essential to all skin fittings on thin hulls (such as metal) to avoid stresses. Fit a generous teak packing chock inside, and glass the whole thing over on a fibreglass hull, for extra security.

Adding to Alloy

With light alloy hulls, even deck and rigging fittings (and their fastenings) must be of compatible metal, for galvanic action occurs slowly even in salt air and spray. Insulation pads of Tufnol are helpful when there is the slightest doubt.

Many wooden craft have alloy masts and spars, so be careful when adding cleats and other fittings to these. Although sound anodizing helps to insulate the alloy, new slides of the wrong metal could cause trouble as the track anodizing gets worn away. Don't use a bronze shackle between anchor and cable. As can be seen from the galvanic table in Chapter 4, not even stainless steel is inert when close to some other metals.

Instruments

When the restless handyman runs short of basic improvements, his sights may turn towards a host of electronic navigational instruments; auto-pilot steering gear, gas detector, wind speed and direction indicator and other useful devices from the chandlery catalogues.

Unfortunately, such items require a deep purse. Some say they rob the hobby a little of its excitement, challenge and fun. Some say they add greatly to safety afloat. Some say they increase the likelihood of vandalism and robbery – but then you can always spend a bit more and buy a burglar alarm!

Index